Nečas Center Series

The Nečas Center Series aims to publish high-quality monographs, textbooks, lecture notes, habilitation and Ph.D. theses in the field of mathematics and related areas in the natural and social sciences and engineering. There is no restriction regarding the topic, although we expect that the main fields will include continuum thermodynamics, solid and fluid mechanics, mixture theory, partial differential equations, numerical mathematics, matrix computations, scientific computing and applications. Emphasis will be placed on viewpoints that bridge disciplines and on connections between apparently different fields. Potential contributors to the series are encouraged to contact the editor-in-chief and the manager of the series.

More information about this series at http://www.springer.com/series/16005

Miroslav Rozložník

Saddle-Point Problems and Their Iterative Solution

 Birkhäuser

Miroslav Rozložník
Institute of Mathematics
Czech Academy of Sciences
Prague, Czech Republic

ISSN 2523-3343   ISSN 2523-3351 (electronic)
Nečas Center Series
ISBN 978-3-030-01430-8   ISBN 978-3-030-01431-5 (eBook)
https://doi.org/10.1007/978-3-030-01431-5

Library of Congress Control Number: 2018962491

This book is published under the imprint Birkhäuser, www.birkhauser-science.com by the registered company Springer Nature Switzerland AG
The registered company address is: Gewerbestrasse 11, 6330 Cham, Switzerland

Preface

It is often rumored that a "saddle point" in mathematics derives its name from the fact that the prototypical example in two dimensions is a surface that curves up in one direction and curves down in a different direction, resembling a riding saddle or a mountain pass between two peaks forming a landform saddle, see Figs. P.1 and P.2. In this contribution, we deal with a solution of linear algebraic systems with a particular 2-by-2 block structure, whereas the lower diagonal block is a zero matrix, and one of the two off-diagonal blocks is the transpose to the other. Since such systems arise also as the first-order optimality conditions in equality-constrained quadratic programming and any of its solutions represents a saddle point in the abovementioned meaning, we use the term "saddle-point problem" for the whole class of such problems. If the (2, 2)-block of this matrix is nonzero, then we use the term "generalized saddle-point problem."

The importance of saddle-point problems or generalized saddle-point problems stems from the fact that they arise in many applications of computational science, engineering, and humanities. Although it is impossible to cover all existing applications that lead to the solution of saddle-point problems and that sometimes naturally overlap, an attempt to keep and to regularly update a collection of various application fields and mathematical disciplines has been made by Michele Benzi (see the schematic list in Fig. P.3).

A lot of attention has been paid to saddle-point problems and their solution in the last two or three decades. A wide variety of results together have appeared in journal articles and conference proceedings as well as in comprehensive surveys and monographs. The excellent survey on numerical solution of saddle-point problems was given by Michele Benzi, Gene Golub, and Jörg Liesen in [12]. Although its purpose was to review the most promising solution approaches for saddle-point problems with an emphasis on iterative methods for large sparse systems, it became the most cited publication in the field, and it still reflects the contemporary state of the art. This fact is clearly visible in the first part of this textbook that loosely follows the subdivision used in [12]. The extensions included here are pointed out in the outline below.

The book by Howard Elman, David Silvester, and Andy Wathen [25] with a particular focus on applications in incompressible fluid dynamics is used not only as the basic reference book in computational fluid dynamics, but it is considered as a fundamental contribution to general numerical linear algebra concepts for solution of saddle-point problems such as the convergence analysis of MINRES, the block-diagonal and block-triangular preconditioning, or the theory of preconditioned iterative methods of $\mathcal{H}$-symmetric saddle-point problems. However, in many applications the information from the origin of saddle-point problems is absolutely essential. Discretization of partial differential equations such as Stokes or Navier-Stokes equations leads often to saddle-point problems. A variety of iterative methods for their solution have been proposed that have rates of convergence independent of the mesh size used in discretization. In most cases, these methods require a preconditioner that is spectrally equivalent or norm equivalent to the system matrix. This is frequently achieved only by very advanced techniques such as multigrid methods that certainly belong to the most efficient methods for solving large discretized problems [9, 21, 24, 48, 49, 68].

The continuous formulation of the original problem leads directly to natural preconditioning that guarantees the fast convergence of iterative methods. This transformation is often called operator preconditioning, and it motivates the construction of practical preconditioners used to accelerate the convergence of iterative methods for solution of resulting discretized problems. The bounds on convergence of iterative methods developed using the norm or spectral equivalence on operator level are then independent of discretization, while traditional approach is based on equivalence of matrices from a particular discretization and a particular preconditioner. These ideas were developed, e.g., in [5, 6, 28, 41, 43], or [49]. Indeed, in the context of numerical solution of partial differential equations, the discretization and efficient preconditioning should be tightly linked due to the fact that a preconditioner can be seen as a transformation of the discretization basis in the finite-dimensional given Hilbert space (see the book by Josef Málek and Zdeněk Strakoš [56]).

As it is seen from previous discussion, saddle-point problems represent a very wide research area with a large amount of work devoted to various applications. In this textbook we focus on some linear algebra aspects of solving saddle-point problems with emphasis on iterative methods, their analysis, implementation, and numerical behavior. We concentrate mainly on algebraic techniques that lead to comprehensive solvers for various saddle-point problems. Nevertheless, each of them can be adapted to particular class of problems with a specific application in mind. A great progress has been made toward an effective preconditioning of iterative methods that in many of such cases leads to very efficient solvers. Although we here briefly discuss some selected applications leading to saddle-point problems, we do not give a detailed treatment of any particular application-based approach. This textbook is based on the course "saddle-point problems and their solution" that is since 2014 included into the education program at the Department of Numerical Mathematics, Charles University in Prague. The course attempts to cover not only classical results on the solution of saddle-point problems that appeared in books, articles, and proceedings, but it also contains the presentation of original results achieved by the author and his colleagues. In particular, we concentrate on numerical behavior issues that have attracted considerably less attention than many other topics related to solving saddle-point problems. We analyze the accuracy of approximate solutions computed by inexact saddle-point solvers, where the solution of certain subproblems is replaced by a cheap relaxation with a relatively modest and easy-to-fulfill requirements. We also look at numerical behavior of certain iterative methods when applied to saddle-point problems with indefinite preconditioning. As an illustration, we consider also the case study with an example from a real-world application. The main idea here is to compare three main solution approaches without any preconditioning that lead to the same asymptotic (but suboptimal) rate of convergence. The development of the solver with optimal convergence rate independent on the discretization parameter that would require to use the information from the underlying continuous problem is out of the scope of this textbook.

The course represents a collection of nine relatively self-contained lectures with separate lists of relevant references. This textbook contains also nine chapters, but the bibliography is extended, unified, and moved toward the end. The first chapter is devoted to introductory remarks on saddle-point matrices and their indefiniteness and to the already mentioned saddle-point motivation from equality-constrained quadratic programming problems and second-order elliptic equations. In Chap. 2, we recall three prominent application fields that lead to saddle-point problems as augmented systems in least squares problems, in the form of linear systems from discretizations of partial differential equations with constraints and as Kuhn-Karush-Tucker systems in interior-point methods. The first part gives some basic facts on least squares methods that are useful also in the context of constraint preconditioning and covers some generalizations leading to saddle-point problems. Instead of a specification of a particular continuous problem, the second part on saddle-point problems that arise from discretization of partial differential equations uses a general abstract framework and formulation as mixed variational problem in certain Hilbert spaces together with the discretization in their finite-dimensional subspaces. The first part of Chap. 3 formulates the necessary and sufficient condition for the saddle-point matrix to be nonsingular and gives a review of basic results on the inverse of a saddle-point matrix that form essentially a background for two main solution approaches. In the second part, a special attention is paid to the spectral properties of this particular class of symmetric indefinite matrices including also the results on their inertia and on eigenvalue inclusion sets. Some of them were developed quite recently as those in the case of a semi-definite diagonal block or a rank-deficient off-diagonal block. Two main solution approaches, the Schur complement method and the null-space method, are discussed in Chap. 4 including their schematic algorithms in the general inner-outer iteration setting. The notion of the Schur complement method and the null-space method reappears several times throughout this textbook. We consider these two approaches in the context of factorization of saddle-point matrices, stationary iterative methods with indefinite splitting matrix, constraint preconditioning, and numerical behavior of inexact saddle-point solvers. Chapter 5 is devoted to the direct solution of saddle-point problems with a focus on the LDL^T factorization of symmetric indefinite matrices. It appears that under standard assumptions, the saddle-point matrix admits such a factorization, there is no need for pivoting. In addition, the condition number of the triangular factor can be explicitly bounded in terms of the condition numbers of the whole saddle-point matrix and diagonal block. Next, two main solution approaches are recalled again from a perspective of the direct solution of saddle-point problems. The central idea of Chap. 6 on the iterative solution of saddle-point problems using stationary iterative methods and Krylov subspace methods is to distinguish between three different cases: solution of the whole saddle-point system, solution of the Schur complement system, and solution of the system projected onto the null-space of the off-diagonal block. Therefore, we briefly discuss the most widely known and used Krylov subspace methods applied to symmetric positive definite systems, symmetric indefinite systems, and nonsymmetric systems: CG, MINRES, and GMRES. In particular, preconditioned Krylov subspace methods are reviewed very carefully treating all relevant combinations of the symmetric positive definite or indefinite system with the symmetric positive definite, symmetric indefinite, or nonsymmetric preconditioner. We briefly cover also multigrid methods that are successfully used for solving saddle-point problems that arise from discretizations of partial differential equations and give links to stationary iterative methods that represent their key ingredient in the form of smoothing procedure. Chapter 7 gives a survey on block preconditioners for saddle-point problems including block-diagonal, block-triangular, and constraint preconditioners. The focus is put on the relation of constraint preconditioners to the Schur complement or null-space method as these can be seen as applications of indefinitely preconditioned Krylov subspace method on the saddle-point problem with a particular initial guess. In Chap. 8 we concentrate on the numerical behavior of the Schur complement reduction method and the null-space method. Without going too much into details and without rigorous proofs of main results, we discuss the effects of inexact solution of inner systems on the maximum attainable accuracy of approximate solutions computed in the Schur complement method and in the null-space method with respect to the back-substitution formula used for computing the other approximate solutions to saddle-point system. We point out the optimal implementations delivering high-accuracy approximate solutions that are used in practical computations. We also study the influence of the scaling of the diagonal block in the saddle-point system solved with the conjugate gradient method and preconditioned with the corresponding constraint preconditioner on the accuracy of approximate solutions computed by these variants of the Schur complement method and of the null-space projection method. The described phenomena occur universally for all problems. In each case they are illustrated on small saddle-point problems, where we can also monitor their conditioning. Finally, Chap. 9 contains the case study that comes from a real-world application of groundwater flow modeling in the area of Stráž pod Ralskem in northern Bohemia. We consider the potential fluid flow in porous media discretized by the mixed-hybrid finite element method using trilateral prismatic elements with a uniformly regular mesh refinement leading to large-scale saddle-point problems with a particular block structure. The convergence behavior of the unpreconditioned MINRES method applied to the whole saddle-point problem, the Schur complement systems, and the systems projected onto certain null-spaces without preconditioning is analyzed using the tools described in previous chapters of this textbook. Since in the general indefinite case with positive and negative eigenvalues it is difficult to get sharp and descriptive bounds using the Chebyshev polynomials as in the positive definite case, for comparison of all approaches, the asymptotic convergence factor is used as a first indicator for describing their convergence behavior. It follows that the bounds for these three solution approaches are comparable in terms of the discretization parameter, and in practical situations they require efficient preconditioning that would ideally lead to convergence rate independent of discretization. The developed results are illustrated on a simple potential fluid flow problem in an artificial cubic domain.

Acknowledgments I would like to thank the coauthors of our joint papers related to the subject of this textbook for their precious collaboration, kind hospitality, and friendship over the years. My thanks go to P. Bastian, J. Málek, V. Mehrmann, M. Pokorný, and Z. Strakoš, the members of the editorial board of Nečas Center Series, for their valuable suggestions that have led to a significant improvement of the manuscript. I am grateful to J. Maryška and J. Mužák who introduced us to the application solved in northern Bohemia and provided us with plots from DIAMO, s. e. in Stráž pod Ralskem. I am indebted to M. Benzi, D. Dvořáková, I. Bohušová, and Ľ. Rusnáková for sending me the collection of applications leading to saddle-point problems and for for allowing me to reprint their beautiful saddles. I would like to thank M. Křížek, J. Kuřátko, and V. Kubáč for careful reading the manuscript and many useful comments. I have greatly benefitted from the help of H. Bílková and O. Ulrych with the peculiarities of the LaTeXdocument preparation system. This publication has been prepared with the support of the Nečas Center for Mathematical Modeling. The author was also supported by the Czech Science Foundation under the Grant 18-09628S "Advanced flow-field analysis" and by institutional support RVO 67985840.

Prague, Czech Republic   Miro(slav) Rozložník
June 2018

Fig. P.1 Bone horse saddles from the fifteenth century, traditionally associated with King Sigismund of Luxembourg (1368–1437), King of Hungary and Croatia (1387–1437), King of Germany (1411–1437), King of Bohemia (1419–1437), King of Italy (1431–1437) and Holy Roman Emperor (1433–1437). Collections of Hungarian National Museum, Budapest. (Courtesy of D. Dvořáková: Kôň a človek v stredoveku, Vydavateľstvo Rak, Budmerice, 2007)

Fig. P.2 Priečne sedlo (saddle) (2352 m above sea level) located in High Tatras, Slovakia, is a narrow pass from Malá Studená dolina to Velká Studená dolina, two beautiful mountain valleys, connecting in fact Téryho chata and Zbojnícka chata cottages. Sedielko ("little saddle") (2376 m above sea level) is the highest tourist pass across the main mountain ridge of High Tatras, connecting Malá Studená dolina with Zadná Javorová dolina. Priečne sedlo and Sedielko are connected with the ridge that summits in the peak Široká veža (2462 m above sea level). (Photo taken by I. Bohuš, ml. Courtesy of I. Bohušová, Vydavateľstvo IB Tatry, Tatranská Lomnica)

Computational-fluid-dynamics
                              Computer-graphics   Image-up-sampling
Tomography   Ocean-climate-modeling
              Magnetostatics   Electromagnetism
                              Finance   Economics   Electrical-circuits
Networks   Thermoelasticity
            Linear-elasticity
Earth-sciences   Image-registration   Shape-and-topology-optimization
Contact-mechanics   Elasticity-fluid-mechanics   Simulation-of-glaciers
                                                  Design-of-bipedal-robots
Electro-encephalography   Image-restoration
                                                           Poroelasticity
Power-system-state-estimation   Non-linear-elasticity
SAW-driven-biochips   Power-plant-flow-simulations   Dictionary-design
Computational-geodynamics   Acoustic-streaming   Electrocardiology
Stokes-problems   Analysis-of-photonic-crystals   Chemical-engineering
Mixed-finite-element-method   RCL-circuit-simulations   Darcy-problems
Oseen-problems   Hydrodynamics-modelling   Navier-Stokes-problems
Constrained-optimization   Game-theory   Weighted-least-squares-problems
Constrained-least-squares-problems   PDE-constrained-optimization
Mesh-generation   Piezoelectric-material-tensors   Interior-point-methods
              Parameter-identification-problems   Mesh-analysis
Interpolation-of-scattered-data   Optimal-control   Kalman-filtering
Model-order-reduction   Structural-and-multidisciplinary-optimization
Semidefinite-programming   Biot's-model   DAE's-in-water-networks
Random-Markov-fields   Statistics   Equilibrium-in-bimatrix-games
Land-mine-signatures   Allen-Cahn-equations   Consolidation-equations
Biphasic-materials-in-biomechanics   Multicommodity-network-flow
Thermal-stress-analysis Image-reconstruction Electric-field-integral-equations
Nanofiber-textile-problems   Stochiometric-matrices   Lyapunov-equations
Optimal-transport-problems   Data-assimilation   Stochastic-programming
Bimatrix-games   Nonlinear-eigenvalue-problems   Systems-analysis
Variational-inequalities   Chemical-transport-reaction-models
Liquid-crystal-directors-modeling   Bidomain-reaction-diffusion-system

Fig. P.3 Collection of applications and mathematical disciplines leading to saddle-point problems. (Courtesy of M. Benzi)

Contents

Chapter 1
Introductory Remarks

Throughout this manuscript we consider the system of linear algebraic equations

$$\mathbb{A}x = \mathbb{b}, \tag{1.1}$$

where $\mathbb{A} \in \mathcal{R}^{m+n,m+n}$ is a real $(m + n)$-by-$(m + n)$ nonsingular matrix with the 2-by-2 block structure

$$\mathbb{A} = \begin{pmatrix} A & B \\ B^T & C \end{pmatrix}, \tag{1.2}$$

where $A \in \mathcal{R}^{m,m}$ is a square matrix of order m, $B \in \mathcal{R}^{m,n}$ is a rectangular matrix with $m \geq n$, and $C \in \mathcal{R}^{n,n}$ is a square matrix of order n. The vector $\mathbb{b} \in \mathcal{R}^{m+n}$ is called the right-hand side vector, and the vector $x_* \in \mathcal{R}^{m+n}$ satisfying (1.1) is called the exact solution.

Even if $\mathbb{A}$ is nonsingular, the diagonal block A or the diagonal block C can be singular; also the off-diagonal block B can be rank-deficient. The examples of such cases are the following:

$$\mathbb{A} = \begin{pmatrix} 0 & 1 \\ 1 & 1 \end{pmatrix}, \ \mathbb{A} = \begin{pmatrix} 1 & 1 \\ 1 & 0 \end{pmatrix}, \ \mathbb{A} = \begin{pmatrix} 0 & 1 \\ 1 & 0 \end{pmatrix}, \ \mathbb{A} = \begin{pmatrix} 1 & 0 \\ 0 & 1 \end{pmatrix}.$$

If A and C are symmetric ($A^T = A$ and $C^T = C$), then $\mathbb{A}$ is symmetric with $\mathbb{A}^T = \mathbb{A}$.

Symmetric positive-definite case We consider also the case $\mathbb{A} = \begin{pmatrix} A & B \\ B^T & C \end{pmatrix}$, where $A \in \mathcal{R}^{m,m}$, $B \in \mathcal{R}^{m,n}$, and $C \in \mathcal{R}^{n,n}$ such that $\mathbb{A}$ is symmetric positive definite with $x^T \mathbb{A} x > 0$ for $x \neq 0$.

© Springer Nature Switzerland AG 2018
M. Rozložník, *Saddle-Point Problems and Their Iterative Solution*,
Nečas Center Series, https://doi.org/10.1007/978-3-030-01431-5_1

It is clear that if $\mathbb{A}$ is symmetric positive definite, then $\mathbb{A}$ is nonsingular. In addition, all principal submatrices of $\mathbb{A}$ are symmetric positive definite and thus nonsingular. Both A and C are symmetric positive definite (see, e.g., [7]). Indeed, we have

$$\begin{pmatrix} x \\ 0 \end{pmatrix}^T \begin{pmatrix} A & B \\ B^T & C \end{pmatrix} \begin{pmatrix} x \\ 0 \end{pmatrix} = x^T A x > 0, \ x \neq 0,$$

$$\begin{pmatrix} 0 \\ y \end{pmatrix}^T \begin{pmatrix} A & B \\ B^T & C \end{pmatrix} \begin{pmatrix} 0 \\ y \end{pmatrix} = y^T C y > 0, \ y \neq 0.$$

Proposition 1.1 *Assume that A is symmetric positive definite and C is symmetric negative definite. Then $\mathbb{A}$ is indefinite with*

$$\begin{pmatrix} x \\ 0 \end{pmatrix}^T \begin{pmatrix} A & B \\ B^T & C \end{pmatrix} \begin{pmatrix} x \\ 0 \end{pmatrix} = x^T A x > 0, \ x \neq 0,$$

$$\begin{pmatrix} 0 \\ y \end{pmatrix}^T \begin{pmatrix} A & B \\ B^T & C \end{pmatrix} \begin{pmatrix} 0 \\ y \end{pmatrix} = y^T C y < 0, \ y \neq 0.$$

Saddle-point matrix We consider the case when $C \in \mathcal{R}^{n,n}$ in (1.2) is the zero matrix ($C = 0$) with

$$\mathbb{A} = \begin{pmatrix} A & B \\ B^T & 0 \end{pmatrix} \tag{1.3}$$

and where $A \in \mathcal{R}^{m,m}$ and $B \in \mathcal{R}^{m,n}$ are such that $\mathbb{A}$ is nonsingular. It is often the case that the right-hand side vector $\mathbb{b}$ has the form

$$\mathbb{b} = \begin{pmatrix} c \\ 0 \end{pmatrix}. \tag{1.4}$$

Assume a saddle-point matrix $\mathbb{A}$ with a symmetric positive definite A and a full-column rank B. Then for any $x \neq 0$ and $y \neq 0$, we have

$$\begin{pmatrix} x \\ 0 \end{pmatrix}^T \begin{pmatrix} A & B \\ B^T & 0 \end{pmatrix} \begin{pmatrix} x \\ 0 \end{pmatrix} = x^T A x > 0,$$

$$\begin{pmatrix} -A^{-1}By \\ y \end{pmatrix}^T \begin{pmatrix} A & B \\ B^T & 0 \end{pmatrix} \begin{pmatrix} -A^{-1}By \\ y \end{pmatrix} = -y^T B^T A^{-1} B y < 0.$$

In the general case of a nonsingular saddle-point matrix $\mathbb{A}$ with a symmetric positive semi-definite A and a full-column rank B, it follows from $\mathcal{N}(A) \cap \mathcal{N}(B^T) = \{0\}$ that for all nonzero $x \in \mathcal{N}(B^T)$, we have $x^T A x > 0$ and therefore

$$\begin{pmatrix} x \\ 0 \end{pmatrix}^T \begin{pmatrix} A & B \\ B^T & 0 \end{pmatrix} \begin{pmatrix} x \\ 0 \end{pmatrix} = x^T A x > 0, \quad x \neq 0, \ x \in \mathcal{N}(B^T).$$

In addition, taking any $y \neq 0$, we get

$$\begin{pmatrix} By \\ 0 \end{pmatrix}^T \begin{pmatrix} A & B \\ B^T & 0 \end{pmatrix} \begin{pmatrix} By \\ 0 \end{pmatrix} = (By)^T A By \geq 0,$$

$$\begin{pmatrix} By \\ -(B^T B)^{-1} B^T A By \end{pmatrix}^T \begin{pmatrix} A & B \\ B^T & 0 \end{pmatrix} \begin{pmatrix} By \\ -(B^T B)^{-1} B^T A By \end{pmatrix} = -(By)^T A By \leq 0.$$

Motivation for saddle-point problems First we consider the equality-constrained quadratic programming problem

$$\text{minimize } f(x) = \frac{1}{2} x^T A x - c^T x \text{ over all } x \in \mathcal{R}^m$$

$$\text{subject to } B^T x = d,$$

where $A \in \mathcal{R}^{m,m}$ is symmetric positive definite, $B \in \mathcal{R}^{m,n}$ is of full-column rank, $c \in \mathcal{R}^m$, and $d \in \mathcal{R}^n$. We define the corresponding Lagrangian function

$$\mathcal{L}(x, y) \equiv f(x) + y^T (B^T x - d) = \frac{1}{2} x^T A x + x^T B y - x^T c - y^T d$$

$$= \frac{1}{2} \begin{pmatrix} x \\ y \end{pmatrix}^T \begin{pmatrix} A & B \\ B^T & 0 \end{pmatrix} \begin{pmatrix} x \\ y \end{pmatrix} - \begin{pmatrix} x \\ y \end{pmatrix}^T \begin{pmatrix} c \\ d \end{pmatrix},$$

where the vector $y \in \mathcal{R}^n$ represents the vector of Lagrange multipliers.

Proposition 1.2 ([62], Chapter 16.1) *Any solution* $\mathbb{x}_* = \begin{pmatrix} x_* \\ y_* \end{pmatrix}$ *of the system* (1.1), *where* $\mathbb{A} = \begin{pmatrix} A & B \\ B^T & 0 \end{pmatrix}$ *and* $\mathbb{b} = \begin{pmatrix} c \\ d \end{pmatrix}$, *is a stationary point of the Lagrangian* $\mathcal{L}(x, y)$.

Proof We denote the elements of matrices $A = (a_{i,j})$ and $B = (b_{i,k})$ for $i = 1, \ldots, m$, $j = 1, \ldots, m$, and $k = 1, \ldots, n$. The elements of vectors $c \in \mathcal{R}^m$ and

$d \in \mathcal{R}^n$ are denoted as $c = (c_1, \ldots, c_m)^T$ and $d = (d_1, \ldots, d_n)^T$. Since

$$\mathcal{L}(x, y) = \sum_{i=1}^{m} x_i \left(\sum_{j=1}^{m} a_{i,j} x_j + \sum_{k=1}^{n} b_{i,k} y_k - c_i \right) - \sum_{k=1}^{n} y_k d_k$$

$$= \sum_{i=1}^{m} x_i \left(\sum_{j=1}^{m} a_{i,j} x_j - c_i \right) + \sum_{k=1}^{n} y_k \left(\sum_{j=1}^{m} b_{k,j}^T x_j - d_k \right)$$

for any $x = (x_1, \ldots, x_m)^T \in \mathcal{R}^m$ and $y = (y_1, \ldots, y_n)^T \in \mathcal{R}^n$, by differentiation we obtain

$$\frac{\partial \mathcal{L}}{\partial x_j}(x, y) = e_j^T (Ax - By - c), \quad j = 1, \ldots, m,$$

$$\frac{\partial \mathcal{L}}{\partial y_k}(x, y) = e_k^T (B^T x - d), \quad k = 1, \ldots, n,$$

where $e_j \in \mathcal{R}^m$ and $e_k \in \mathcal{R}^n$ denote the j-th and k-th column of identity matrices of corresponding dimensions, respectively. Thus the first-order optimality conditions then read as follows

$$\begin{pmatrix} A & B \\ B^T & 0 \end{pmatrix} \begin{pmatrix} x \\ y \end{pmatrix} - \begin{pmatrix} c \\ d \end{pmatrix} = \begin{pmatrix} 0 \\ 0 \end{pmatrix}. \tag{1.5}$$

$\square$

For any vector $\mathbb{x} = \begin{pmatrix} x \\ y \end{pmatrix} \in \mathcal{R}^{m+n}$ and for any vector $\mathbb{x}_* = \begin{pmatrix} x_* \\ y_* \end{pmatrix} \in \mathcal{R}^{m+n}$ satisfying (1.5), we have

$$\mathcal{L}(x_*, y_*) = \frac{1}{2} \begin{pmatrix} x_* \\ y_* \end{pmatrix}^T \begin{pmatrix} A & B \\ B^T & 0 \end{pmatrix} \begin{pmatrix} x_* \\ y_* \end{pmatrix} - \begin{pmatrix} x_* \\ y_* \end{pmatrix}^T \begin{pmatrix} c \\ d \end{pmatrix}$$

$$= -\frac{1}{2} \begin{pmatrix} x_* \\ y_* \end{pmatrix}^T \begin{pmatrix} A & B \\ B^T & 0 \end{pmatrix} \begin{pmatrix} x_* \\ y_* \end{pmatrix},$$

$$\mathcal{L}(x, y) = \frac{1}{2} \begin{pmatrix} x \\ y \end{pmatrix}^T \begin{pmatrix} A & B \\ B^T & 0 \end{pmatrix} \begin{pmatrix} x \\ y \end{pmatrix} - \begin{pmatrix} x \\ y \end{pmatrix}^T \begin{pmatrix} A & B \\ B^T & 0 \end{pmatrix} \begin{pmatrix} x_* \\ y_* \end{pmatrix}.$$

Indeed it holds for any $x \in \mathcal{R}^m$ and for any $y \in \mathcal{R}^n$

$$\mathcal{L}(x, y) - \mathcal{L}(x_*, y_*) = \frac{1}{2} \left[\begin{pmatrix} x \\ y \end{pmatrix} - \begin{pmatrix} x_* \\ y_* \end{pmatrix} \right]^T \begin{pmatrix} A & B \\ B^T & 0 \end{pmatrix} \left[\begin{pmatrix} x \\ y \end{pmatrix} - \begin{pmatrix} x_* \\ y_* \end{pmatrix} \right].$$

Consequently, substituting for $y = y_*$, we obtain

$$\mathcal{L}(x, y_*) - \mathcal{L}(x_*, y_*) = \frac{1}{2}(x - x_*)^T A (x - x_*) \geq 0,$$

and substituting for $x = x_*$, we get $\mathcal{L}(x_*, y) - \mathcal{L}(x_*, y_*) = \frac{1}{2}(y - y_*)^T 0 (y - y_*) = 0$. This leads to the basic inequalities

$$\mathcal{L}(x_*, y) \leq \mathcal{L}(x_*, y_*) \leq \mathcal{L}(x, y_*), \quad x \in \mathcal{R}^m, \ y \in \mathcal{R}^n.$$

Proposition 1.3 ([33], Proposition 2.1) *If* $\mathrm{x}_* = \begin{pmatrix} x_* \\ y_* \end{pmatrix}$ *is a saddle point of the Lagrangian* $\mathcal{L}(x, y)$, *then*

$$\min_{x \in \mathbb{R}^m} \max_{y \in \mathbb{R}^n} \mathcal{L}(x, y) = \mathcal{L}(x_*, y_*) = \max_{y \in \mathbb{R}^n} \min_{x \in \mathbb{R}^m} \mathcal{L}(x, y). \tag{1.6}$$

Proof Indeed $\inf_{x \in \mathbb{R}^m} \mathcal{L}(x, y) \leq \mathcal{L}(x, y) \leq \sup_{y \in \mathbb{R}^n} \mathcal{L}(x, y)$. Therefore, we have $\sup_{y \in \mathbb{R}^n} \inf_{x \in \mathbb{R}^m} \mathcal{L}(x, y) \leq \sup_{y \in \mathbb{R}^n} \mathcal{L}(x, y)$ for all $x \in \mathbb{R}^m$. Taking the inf over $\mathbb{R}^m$ of the above expression leads to

$$\sup_{y \in \mathbb{R}^n} \inf_{x \in \mathbb{R}^m} \mathcal{L}(x, y) \leq \inf_{x \in \mathbb{R}^m} \sup_{y \in \mathbb{R}^n} \mathcal{L}(x, y). \tag{1.7}$$

The definition of the saddle-point $\mathrm{x}_* = \begin{pmatrix} x_* \\ y_* \end{pmatrix}$ yields

$$\mathcal{L}(x_*, y) \leq \mathcal{L}(x_*, y_*) \leq \mathcal{L}(x, y_*), \quad x \in \mathcal{R}^m, \ y \in \mathcal{R}^n.$$

Taking the sup over $\mathcal{R}^n$ in the lower bound and the inf over $\mathcal{R}^m$ in the upper bound, respectively, gives $\sup_{y \in \mathbb{R}^n} \mathcal{L}(x_*, y) \leq \mathcal{L}(x_*, y_*) \leq \inf_{x \in \mathbb{R}^m} \mathcal{L}(x, y_*)$. Thus we get

$$\inf_{x \in \mathbb{R}^m} \sup_{y \in \mathbb{R}^n} \mathcal{L}(x, y) \leq \sup_{y \in \mathbb{R}^n} \mathcal{L}(x_*, y)$$

$$\leq \mathcal{L}(x_*, y_*) \leq$$

$$\inf_{x \in \mathbb{R}^m} \mathcal{L}(x, y_*) \leq \sup_{y \in \mathbb{R}^n} \inf_{x \in \mathbb{R}^m} \mathcal{L}(x, y).$$

It follows then from (1.7) that $\inf_{x \in \mathbb{R}^m} \sup_{y \in \mathbb{R}^n} \mathcal{L}(x, y) = \sup_{y \in \mathbb{R}^n} \mathcal{L}(x_*, y) = \mathcal{L}(x_*, y_*)$ $= \inf_{x \in \mathbb{R}^m} \mathcal{L}(x, y_*) = \sup_{y \in \mathbb{R}^n} \inf_{x \in \mathbb{R}^m} \mathcal{L}(x, y).$ $\square$

Another important class of problems in many applications involves the second-order elliptic equation of the form

$$-\nabla \cdot (A \nabla p) = f \text{ in } \Omega, \tag{1.8}$$

where Ω is a bounded domain in $\mathcal{R}^3$ with a Lipschitz continuous boundary $\partial\Omega$, $f \in L^2(\Omega)$ and A is a symmetric and uniformly positive definite second-rank tensor with $[\mathsf{A}(\mathbf{x})]_{ij} \in L^\infty(\Omega)$ for all $i, j \in \{1, 2, 3\}$. The boundary conditions are given by

$$p = p_D \text{ on } \partial\Omega_D, \quad -\mathsf{A}\nabla p \cdot \mathbf{n} = u_N \text{ on } \partial\Omega_N, \tag{1.9}$$

where $\partial\Omega_D$ and $\partial\Omega_N$ are sufficiently regular portions of $\partial\Omega$ such that $\partial\Omega = \overline{\partial\Omega_D} \cup \overline{\partial\Omega_N}$, $\partial\Omega_D \cap \partial\Omega_N = \emptyset$, and $|\partial\Omega_D| > 0$. The vector $\mathbf{n}$ is the outward normal vector defined (almost everywhere) on the boundary $\partial\Omega$, $p_D \in L^2(\partial\Omega_D)$, and $u_N \in L^2(\partial\Omega_N)$. By $L^2(\Omega)$ we here denote the Lebesgue space defined as $L^2(\Omega) = \{\varphi : \Omega \to \mathcal{R} \mid \int_\Omega |\varphi|^2 d\mathbf{x} < \infty\}$ with the inner product $(\varphi, \psi)_\Omega = \int_\Omega \varphi\psi d\mathbf{x}$. We also introduce the bilinear form $(\varphi, \psi)_{\partial\Omega} = \int_{\partial\Omega} \varphi\psi dS$, where φ and ψ are functions from $L^2(\partial\Omega)$. By $H^1(\Omega)$ we denote the Sobolev space defined as $H^1(\Omega) = \{\varphi \in L^2(\Omega) \mid \nabla\varphi \in L^2(\Omega)\}$, and by $H(\text{div}; \Omega)$ we then denote the Sobolev space defined as $H(\text{div}; \Omega) = \{\mathbf{u} \in \mathbf{L}^2(\Omega) \mid \nabla \cdot \mathbf{u} \in L^2(\Omega)\}$.

It was shown in Sect. 6.1 of [69] that there exists a function p_* that is the unique solution of (1.8) minimizing the functional

$$\mathcal{J}(p) = \frac{1}{2}(\mathsf{A}\nabla p, \nabla p)_\Omega - (f, p)_\Omega - (u_N, p)_{\partial\Omega_N} \tag{1.10}$$

over all functions p in the affine space

$$p \in \left\{ p \in H^1(\Omega) \mid p = p_D \text{ on } \partial\Omega_D \right\}. \tag{1.11}$$

Let us now define the vector function $\mathbf{u}_* = -\mathsf{A}\nabla p_*$, where p_* is the solution of (1.8). Then the problem (1.8) can be written as the first-order system with respect to the unknowns $(\mathbf{u}, p)$ in the form

$$\mathbf{u} = -\mathsf{A}\nabla p, \quad \nabla \cdot \mathbf{u} = f \text{ on } \Omega, \tag{1.12}$$

$$p = p_D \text{ on } \partial\Omega_D, \quad \mathbf{u} \cdot \mathbf{n} = u_N \text{ on } \partial\Omega_N. \tag{1.13}$$

It was shown in Sect. 7.1 of [69] that the solution $\mathbf{u}_*$ to (1.12) is given by the minimization of the functional

$$\mathcal{I}(\mathbf{u}) = \frac{1}{2}\left(\mathsf{A}^{-1}\mathbf{u}, \mathbf{u}\right)_\Omega - (\mathbf{u} \cdot \mathbf{n}, p_D)_{\partial\Omega_D} \tag{1.14}$$

over all functions $\mathbf{u}$ in the vector space

$$\{\mathbf{u} \in H(\text{div}; \Omega) \mid \nabla \cdot \mathbf{u} = f, \mathbf{u} \cdot \mathbf{n} = u_N \text{ on } \partial\Omega_N\}. \tag{1.15}$$

In a variety of problems, it is desirable to obtain an approximation to $\mathbf{u}_*$ that fulfills sufficiently well the continuity equation $\nabla \cdot \mathbf{u} = f$ in (1.12). Very often such an accurate approximation can be determined by the mixed method. The idea of mixed method is based on the introduction of a Lagrange multiplier to relax the constraint $\nabla \cdot \mathbf{u} = f$. Introducing the Lagrangian

$$\mathcal{L}(\mathbf{u}, p) = \mathcal{I}(\mathbf{u}) + (\nabla \cdot \mathbf{u} - f, p)_\Omega, \tag{1.16}$$

we look for the saddle point $(\mathbf{u}_*, p_*)$ that minimizes $\mathcal{L}(\mathbf{u}, p)$ over $\mathbf{u}$ and maximizes $\mathcal{L}(\mathbf{u}, p)$ over p, i.e.,

$$\mathcal{L}(\mathbf{u}_*, p) \leq \mathcal{L}(\mathbf{u}_*, p_*) \leq \mathcal{L}(\mathbf{u}, p_*). \tag{1.17}$$

The saddle-point problem (1.17) has a unique solution $(\mathbf{u}_*, p_*)$ in

$$\{\mathbf{u} \in H(\text{div}; \Omega) \mid \mathbf{u} \cdot \mathbf{n} = u_N \text{ on } \partial\Omega_N\} \times L^2(\Omega),$$

whereas p_* is the solution of (1.8) and $\mathbf{u}_*$ that satisfies $\nabla \cdot \mathbf{u} = f$ is related to p_* through $\mathbf{u}_* = -\mathbf{A}\nabla p_*$. For a detailed treatment of mixed methods for elliptic problems that are based on relaxation of constraints via a saddle point, we refer to Chap. 7 in [69]; see also [17].

Chapter 2
Selected Applications Leading to Saddle-Point Problems

2.1 Augmented Systems in Least Squares Problems

In the following we consider the problem of finding a vector $y \in \mathcal{R}^n$ such that $By \approx c$, where $B \in \mathcal{R}^{m,n}$ with $m \geq n$ is the data matrix and $c \in \mathcal{R}^m$ is the observation vector. If $m > n$, there are more equations than unknowns in the overdetermined linear system $By = c$. In general, its classical solution may not exist, and therefore, we consider the following generalization.

Definition 2.1 Given $B \in \mathcal{R}^{m,n}$ and $c \in \mathcal{R}^m$, we consider the problem

$$\|c - By_*\| = \min_{y \in \mathcal{R}^n} \|c - By\|, \tag{2.1}$$

where $\| \cdot \|$ denotes the standard Euclidean norm. We call this problem a linear least squares problem , and the vector y_* is a linear least squares solution.

If $\operatorname{rank}(B) < n$, then the solution of (2.1) is not unique. However, among all least squares solutions satisfying (2.1), there exists a unique minimum norm solution that minimizes the norm $\|y_*\|$.

The range-space (or the column-space) of the matrix $B \in \mathcal{R}^{m,n}$ is defined as $\mathcal{R}(B) = \{x \in \mathcal{R}^m \mid x = By, \ y \in \mathcal{R}^n\}$. The set of solutions to $B^T x = 0$ is a subspace called the null-space (or the kernel) of B^T, i.e., $\mathcal{N}(B^T) = \{x \in \mathcal{R}^m \mid B^T x = 0\}$, and it represents the orthogonal complement to the space $\mathcal{R}(B)$ in $\mathcal{R}^m$, i.e., $\mathcal{R}(B) \oplus \mathcal{N}(B^T) = \mathcal{R}^m$ with $\mathcal{N}(B^T) = (\mathcal{R}(B))^\perp$. Similarly, one can show that $\mathcal{R}(B^T) \oplus \mathcal{N}(B) = \mathcal{R}^n$ with $\mathcal{N}(B) = (\mathcal{R}(B^T))^\perp$.

We will show now that any least square solution y_* uniquely decomposes the right-hand side c into two orthogonal components

$$c = By_* + c - By_*, \quad By_* \in \mathcal{R}(B), \quad c - By_* \in \mathcal{N}(B^T).$$

© Springer Nature Switzerland AG 2018
M. Rozložník, *Saddle-Point Problems and Their Iterative Solution*,
Nečas Center Series, https://doi.org/10.1007/978-3-030-01431-5_2

Proposition 2.1 *The vector $y_* \in \mathcal{R}^n$ is a linear least squares solution $\iff$ it solves the system of normal equations*

$$B^T B y = B^T c, \tag{2.2}$$

i.e., the residual vector $c - By_*$ belongs to the null-space $\mathcal{N}(B^T)$.

Proof The system of normal equations (2.2) is always consistent, since $B^T c \in \mathcal{R}(B^T) = \mathcal{R}(B^T B)$. The inclusion $\mathcal{R}(B^T B) \subset \mathcal{R}(B^T)$ is trivial. If we take some $y \in \mathcal{R}(B^T)$, then there exists some $x \in \mathcal{R}^m$ such that $y = B^T x$. Decomposing x uniquely into $x = B\tilde{y} + \tilde{x}$, where $B^T \tilde{x} = 0$, we get that $y = B^T(B\tilde{y} + \tilde{x}) = B^T B\tilde{y} \in \mathcal{R}(B^T B)$. This gives the converse inclusion $\mathcal{R}(B^T) \subset \mathcal{R}(B^T B)$.
$\Leftarrow$: Assume that $y_* \in \mathcal{R}^n$ satisfies $B^T(c - By_*) = 0$, i.e., $c - By_* \in \mathcal{N}(B^T)$. Then for any $y \in \mathcal{R}^n$, we have

$$\begin{aligned}
\|c - By\|^2 &= \|c - By_* + B(y_* - y)\|^2 \\
&= \|c - By_*\|^2 + 2\left(B^T(c - By_*), y_* - y\right) + \|B(y_* - y)\|^2 \\
&= \|c - By_*\|^2 + \|B(y_* - y)\|^2.
\end{aligned}$$

The norm of $c - By$ is minimized if $y_* - y \in \mathcal{N}(B)$.
$\Rightarrow$: Suppose that $B^T(c - By_*) \neq 0$ and take $y = y_* + \varepsilon B^T(c - By_*)$, where $\varepsilon > 0$. Then $c - By = c - By_* - \varepsilon B B^T(c - By_*)$ and

$$\begin{aligned}
\|c - By\|^2 &= \|c - By_*\|^2 - 2\varepsilon \|B^T(c - By_*)\|^2 + \varepsilon^2 \|BB^T(c - By_*)\|^2 \\
&< \|c - By_*\|^2
\end{aligned}$$

for sufficiently small ε. This is in contradiction with (2.1). $\qquad\square$

Proposition 2.2 *The matrix $B^T B \in \mathcal{R}^{n,n}$ is symmetric and positive semi-definite. The matrix $B^T B$ is positive definite if and only if the columns of B are linearly independent, i.e., $\mathrm{rank}(B) = n$.*

Proof It is clear that for any $y \in \mathcal{R}^n$, we have that $y^T B^T B y = \|By\|^2 \geq 0$.
$\Leftarrow$: If $\mathrm{rank}(B) = n$, then $y \neq 0$ implies $By \neq 0$, and therefore, $y^T B^T B y = \|By\|^2 > 0$. Hence, $B^T B$ is positive definite.
$\Rightarrow$: If the columns of B are linearly dependent, then for some $y \neq 0$, we have $By = 0$. Consequently, $\|By\|^2 = y^T B^T B y = 0$ and $B^T B$ is not positive definite.
$\qquad\square$

If $\mathrm{rank}(B) < n$, then B has a nontrivial null-space $\mathcal{N}(B)$, and the least squares solution of (2.1) is not unique. If $\tilde{y}_*$ is a particular least squares solution, then any least squares solution y_* is of the form $y_* \in \tilde{y}_* + \mathcal{N}(B)$ leading to $y_* - \tilde{y}_* \in \mathcal{N}(B)$. If, in addition, $\tilde{y}_* \perp \mathcal{N}(B)$, i.e., $\tilde{y}_* \in \mathcal{R}(B^T)$, then

$$\begin{aligned}
\|y_*\|^2 &= \|\tilde{y}_*\|^2 + 2(\tilde{y}_*, y_* - \tilde{y}_*) + \|y_* - \tilde{y}_*\|^2 \\
&= \|\tilde{y}_*\|^2 + \|y_* - \tilde{y}_*\|^2.
\end{aligned}$$

Therefore, $\tilde{y}_*$ is the unique minimum norm least squares solution of (2.1). □

Definition 2.2 Given $B \in \mathcal{R}^{m,n}$ and $d \in \mathcal{R}^n$, we consider the problem of computing the minimum norm solution $x_* \in \mathcal{R}^m$ to the underdetermined system of linear equations $B^T x = d$ in the form

$$\|x_*\| = \min_{x \in \mathcal{R}^m} \|x\| \text{ subject to } B^T x = d. \tag{2.3}$$

If $\text{rank}(B) = n$, then $B^T x = d$ is consistent, and the unique solution x_* of this problem is given in the form $x_* = By_*$, where the vector y_* solves the normal equations of the second kind

$$B^T B y = d. \tag{2.4}$$

Proposition 2.3 (Augmented system formulation) *Assume that* $B \in \mathcal{R}^{m,n}$ *has a full-column rank* $(\text{rank}(B) = n)$*. Then the symmetric* $(m + n)$*-by-*$(m + n)$ *linear system (often called the augmented system)*

$$\begin{pmatrix} I & B \\ B^T & 0 \end{pmatrix} \begin{pmatrix} x \\ y \end{pmatrix} = \begin{pmatrix} c \\ d \end{pmatrix} \tag{2.5}$$

is nonsingular and gives the solutions of the following two problems

$$\min_{y \in \mathcal{R}^n} \{\|c - By\|^2 + 2d^T y\}, \tag{2.6}$$

$$\min_{x \in \mathcal{R}^m} \|x - c\|^2 \text{ subject to } B^T x = d. \tag{2.7}$$

Proof The augmented system (2.5) can be obtained by differentiating the argument $y \in \mathcal{R}^n$ in the problem (2.6) to get $B^T (c - By) = d$ and setting the residual vector $x \in \mathcal{R}^m$ as $x = c - By$. The same system (2.5) can be obtained by differentiating the Lagrangian $\mathcal{L}(x, y) = \|x - c\|^2 + 2y^T (B^T x - d)$ in the problem (2.7) with respect to the argument $x \in \mathcal{R}^m$, where $y \in \mathcal{R}^n$ represents now the vector of Lagrange multipliers. □

Setting $d = 0$, we get the linear least squares problem (2.1), and setting $c = 0$, we get the minimum norm problem (2.3).

Constrained least squares problem Given the matrices $B \in \mathcal{R}^{m,n}$ and $C \in \mathcal{R}^{k,n}$ and the vectors $c \in \mathcal{R}^m$ and $d \in \mathcal{R}^k$, we consider the problem of computing the vector $y_* \in \mathcal{R}^n$ which solves

$$\|c - By_*\| = \min_{y \in \mathcal{R}^n} \|c - By\| \text{ subject to } Cy = d. \tag{2.8}$$

The first-order optimality conditions for this problem read as

$$
\begin{pmatrix} I & 0 & B \\ 0 & 0 & C \\ B^T & C^T & 0 \end{pmatrix} \begin{pmatrix} x \\ z \\ y \end{pmatrix} = \begin{pmatrix} c \\ d \\ 0 \end{pmatrix},
\tag{2.9}
$$

where $z \in \mathcal{R}^k$ is the vector of associated Lagrange multipliers.

Generalized least squares problem In the case when $A \in \mathcal{R}^{m,m}$ is symmetric positive definite, $B \in \mathcal{R}^{m,n}$ has a full-column rank, and $c \in \mathcal{R}^m$, then its inverse A^{-1} is also symmetric positive definite, and we can consider the problem

$$
\|c - By_*\|_{A^{-1}} = \min_{y \in \mathcal{R}^n} \|c - By\|_{A^{-1}},
\tag{2.10}
$$

where $\|\cdot\|_{A^{-1}} = \sqrt{(\cdot, A^{-1}\cdot)}$. The solution of (2.10) can be obtained from the normal equations

$$
B^T A^{-1} By = B^T A^{-1} c.
\tag{2.11}
$$

Introducing the vector $x \in \mathcal{R}^m$ as $x = A^{-1}(c - By)$, the normal equations can be shown to be equivalent to the saddle-point problem in the form

$$
\begin{pmatrix} A & B \\ B^T & 0 \end{pmatrix} \begin{pmatrix} x \\ y \end{pmatrix} = \begin{pmatrix} c \\ 0 \end{pmatrix}.
\tag{2.12}
$$

Since A is symmetric positive definite, it has a Cholesky factorization $A = LL^T$, where $L \in \mathcal{R}^{m,m}$ is nonsingular and lower triangular. Then the generalized least squares problem (2.10) is equivalent to the weighted least squares problem

$$
\|L^{-1}(c - By_*)\| = \min_{y \in \mathcal{R}^n} \|L^{-1}(c - By)\|.
\tag{2.13}
$$

A comprehensive survey of problems and numerical methods for least squares computation can be found in the book [14]; see also Chaps. 1 and 2 of the book [63].

2.2 Saddle-Point Problems from Discretization of Partial Differential Equations with Constraints

Saddle-point problems arise extensively in the numerical analysis of partial differential equations with constraints such as advection-diffusion problem and Stokes and Darcy problems. To address such continuous problems, they are often formulated as mixed variational problems in certain Hilbert spaces, and the mixed finite element method is used for their discretization in the suitably chosen finite-dimensional subspaces leading to large-scale systems of linear algebraic equations.

Let X and Y be two Hilbert spaces equipped with the scalar products $(\cdot, \cdot)_X$ and $(\cdot, \cdot)_Y$, respectively. Their dual spaces consisting of all linear functionals on X and Y are denoted as $X^\#$ and $Y^\#$, respectively. We introduce two bilinear forms $a : X \times X \to \mathcal{R}$ and $b : X \times Y \to \mathcal{R}$ and two linear functionals $c : X \to \mathcal{R}$ and $d : Y \to \mathcal{R}$. We assume that a is symmetric with $a(x, \tilde{x}) = a(\tilde{x}, x)$ and nonnegative with $a(x, x) \geq 0$ for all $x \in X$ and $\tilde{x} \in X$.

We consider the following problem: Find $(x, y) \in X \times Y$ such that

$$a(x, \tilde{x}) + b(\tilde{x}, y) = c(\tilde{x}) \text{ for all } \tilde{x} \in X, \tag{2.14}$$

$$b(x, \tilde{y}) = d(\tilde{y}) \text{ for all } \tilde{y} \in Y. \tag{2.15}$$

If we denote the associated operators $\mathcal{A} : X \to X^\#$ and $\mathcal{B} : X \to Y^\#$ defined as $\langle \mathcal{A}x, \tilde{x} \rangle = a(x, \tilde{x})$ for all $(x, \tilde{x}) \in X \times X$, and $\langle \mathcal{B}x, \tilde{y} \rangle = b(x, \tilde{y})$ for all $(x, \tilde{y}) \in X \times Y$ and the dual operator of $\mathcal{B}$ as $\mathcal{B}^\# : Y \to X^\#$ satisfying $\langle \mathcal{B}^\# y, \tilde{x} \rangle = b(\tilde{x}, y) = \langle \mathcal{B}\tilde{x}, y \rangle$ for $(\tilde{x}, y) \in X \times Y$, then the problem can be reformulated as follows: find $(x, y) \in X \times Y$ such that

$$\mathcal{A}x + \mathcal{B}^\# y = c,$$

$$\mathcal{B}x = d,$$

where $c \in X^\#$ and $d \in Y^\#$ are such that $c(\tilde{x}) = \langle c, \tilde{x} \rangle$ for all $\tilde{x} \in X$ and $d(\tilde{y}) = \langle d, \tilde{y} \rangle$ for all $\tilde{y} \in Y$.

We assume here that both forms a and b are bounded, and therefore there exist positive constants $\alpha_{\max}(\mathcal{A}) > 0$ and $\beta_{\max}(\mathcal{B}) > 0$ such that

$$a(x, x) \leq \alpha_{\max}(\mathcal{A}) \|x\|_X^2, \quad b(x, y) \leq \beta_{\max}(\mathcal{B}) \|x\|_X \|y\|_Y.$$

In addition, we assume that a is coercive on $\mathcal{N}(\mathcal{B})$, i.e.,

$$a(x, x) \geq \alpha_{\min}(\mathcal{A}) \|x\|_X^2 \text{ for all } x \in \mathcal{N}(\mathcal{B}) \tag{2.16}$$

holds for some positive constant $\alpha_{\min}(\mathcal{A}) > 0$.

Definition 2.3 The property of existence of a positive constant $\beta_{\min}(\mathcal{B}) > 0$ such that

$$\inf_{y \in Y} \sup_{x \in X} \frac{b(x, y)}{\|x\|_X \|y\|_Y} \geq \beta_{\min}(\mathcal{B}) \tag{2.17}$$

is known as the inf-sup condition, or the Babuška-Brezzi condition, or the LBB condition (Ladyzhenskaya-Babuška-Brezzi). Then it follows that

$$\sup_{x \in X} \frac{b(x, y)}{\|x\|_X} \geq \beta_{\min}(\mathcal{B}) \|y\|_Y \text{ for all } y \in Y.$$

Theorem 2.1 ([33], Theorem 2.7) *Assume that a is symmetric and coercive on* $\mathcal{N}(\mathcal{B})$*, and assume that the inf-sup condition holds on* $X \times Y$*. The problem* (2.14) *and* (2.15) *has a unique solution. For further details we refer to the lecture notes of the course on the finite element method for fluid mechanics* [33] *and to the paper* [86].

Finite-dimensional approximation Given two finite-dimensional subspaces $X_h \subset X$ and $Y_h \subset Y$, we consider the mixed Galerkin approximation of the continuous problem as follows: Find $(x_h, y_h) \in X_h \times Y_h$ such that

$$a(x_h, \tilde{x}_h) + b(\tilde{x}_h, y_h) = c(\tilde{x}_h) \text{ for all } \tilde{x}_h \in X_h, \tag{2.18}$$

$$b(x_h, \tilde{y}_h) = d(\tilde{y}_h) \text{ for all } \tilde{y}_h \in Y_h. \tag{2.19}$$

The associated operators $\mathcal{A}_h : X_h \to X_h^{\#}$ and $\mathcal{B}_h : X_h \to Y_h^{\#}$ are defined by $\langle \mathcal{A}_h x_h, \tilde{x}_h \rangle = a(x_h, \tilde{x}_h)$ for all $(x_h, \tilde{x}_h) \in X_h \times X_h$ and $\langle \mathcal{B}_h x_h, \tilde{y}_h \rangle = b(x_h, \tilde{y}_h)$ for all $(x_h, \tilde{y}_h) \in X_h \times Y_h$, and the dual operator of B_h as $\mathcal{B}_h^{\#} : Y_h \to X_h^{\#}$ satisfies $\langle \mathcal{B}_h^{\#} y_h, \tilde{x}_h \rangle = b(\tilde{x}_h, y_h) = \langle \mathcal{B}_h \tilde{x}_h, y_h \rangle$ for $(\tilde{x}_h, y_h) \in X_h \times Y_h$.

If we consider the bases $\psi_i, i = 1, \ldots, m$, of the space X_h and $\varphi_k, k = 1, \ldots, n$, of the space Y_h, then any element $x_h \in X_h$ and $y_h \in Y_h$ can be expressed in terms of these bases as

$$x_h = \sum_{i=1}^{m} x_i \psi_i, \quad y_h = \sum_{k=1}^{n} y_k \varphi_k.$$

We define the vectors $c \in \mathcal{R}^m$ as $c = (c(\psi_1), \ldots, c(\psi_m))^T$ and the vector $d \in \mathcal{R}^n$ as $d = (d(\varphi_1), \ldots, d(\varphi_n))^T$. Further, we define the matrices $A \in \mathcal{R}^{m,m}$ and $B \in \mathcal{R}^{m,n}$ as $A = (a_{i,j})$ and $B = (b_{i,k})$, respectively, where $a_{i,j} = a(\psi_i, \psi_j)$ and $b_{i,k} = b(\psi_i, \varphi_k)$ for $i, j = 1, \ldots, m$ and $k = 1, \ldots, n$. Then the system of linear algebraic equations (2.18) and (2.19) takes the saddle-point form

$$\begin{pmatrix} A & B \\ B^T & 0 \end{pmatrix} \begin{pmatrix} x \\ y \end{pmatrix} = \begin{pmatrix} c \\ d \end{pmatrix}. \tag{2.20}$$

Proposition 2.4 ([33], Proposition 3.1) *Assume that a is symmetric and coercive on* $X_h \times X_h$ *and that the inf-sup condition holds on* $X_h \times Y_h$*. Then the matrix A is symmetric positive definite, and the matrix B has the full-column rank.*

Proof Given the bases ψ_i, $i = 1, \ldots, m$, of the space X_h and φ_k, $k = 1, \ldots, n$, of the space Y_h, respectively, we define the vector norm in $\mathcal{R}^m$ of the vector $x = (x_1, \ldots, x_m)^T \in \mathcal{R}^m$ as $\|x\|_X = \|x_h\|_X = \|\sum_{i=1}^m x_i \psi_i\|_X$ and the vector norm in $\mathcal{R}^n$ of the vector $y = (y_1, \ldots, y_n)^T \in \mathcal{R}^n$ as $\|y\|_Y = \|y_h\|_Y = \|\sum_{k=1}^n y_k \varphi_k\|_Y$.
We also define the dual norm in $\mathcal{R}^m$ as $\|x\|_{X^\#} = \sup_{0 \neq \tilde{x} \in \mathcal{R}^m} \frac{|x^T \tilde{x}|}{\|\tilde{x}\|_X}$.

The fact that the matrix A is symmetric positive definite is an immediate consequence of the symmetry and coercivity of a on $X_h \times X_h$. Indeed, the coercivity (and the boundedness) of the bilinear form a reads as follows

$$\alpha_{\min}(A)\|x_h\|_X^2 \leq a(x_h, x_h) \leq \alpha_{\max}(A)\|x_h\|_X^2 \text{ for all } x_h \in X_h,$$

where $0 < \alpha_{\min}(A) \leq \alpha_{\max}(A)$. Then, considering any function $x_h = \sum_{i=1}^m x_i \psi_i$ in X_h and the definition of the matrix $A = (a_{i,j})$ with $a_{i,j} = a(\psi_i, \psi_j)$ for $i, j = 1, \ldots, m$, we have

$$\alpha_{\min}(A)\|x\|_X^2 \leq x^T A x \leq \alpha_{\max}(A)\|x\|_X^2 \text{ for all } x \in \mathcal{R}^m. \tag{2.21}$$

that implies also $x^T A x > 0$ for any $x \neq 0$.

The inf-sup condition on $X_h \times Y_h$ reads as follows

$$\beta_{\min}(B) \leq \inf_{y_h \in Y_h} \sup_{x_h \in X_h} \frac{b(x_h, y_h)}{\|x_h\|_X \|y_h\|_Y}.$$

Taking into account that b is also bounded, there exist positive constants $0 < \beta_{\min}(B) \leq \beta_{\max}(B)$ such that

$$\beta_{\min}(B)\|y_h\|_Y \leq \sup_{x_h \in X_h} \frac{b(x_h, y_h)}{\|x_h\|_X} \leq \beta_{\max}(B)\|y_h\|_Y \quad \text{for all } y_h \in Y_h.$$

These inequalities can be rewritten considering any function $y_h = \sum_{k=1}^n y_k \varphi_k$ in Y_h and the definition of $B = (b_{i,k})$ with $b_{i,k} = b(\psi_i, \varphi_k)$ (where $i = 1, \ldots, m$ and $k = 1, \ldots, n$) into inequalities for vectors y in $\mathcal{R}^n$

$$\beta_{\min}(B)\|y\|_Y \leq \sup_{x \in \mathcal{R}^m} \frac{(x, By)}{\|x\|_X} = \|By\|_{X^\#} \leq \beta_{\max}(B)\|y\|_Y \quad \text{for all } y \in \mathcal{R}^n. \tag{2.22}$$

Indeed, taking the first inequality, $By = 0$ implies $y = 0$, and thus B has a full-column rank. □

The following theorem gives a characterization of eigenvalues of symmetric matrices. It can be used as the starting point for estimates of the extremal eigenvalues of the matrix A.

Theorem 2.2 (min-max theorem, [42]) *If $A \in \mathcal{R}^{m,m}$ is symmetric and positive definite, then*

$$\lambda_{\min}(A) = \min_{0 \neq x \in \mathcal{R}^m} \frac{x^T A x}{x^T x}, \quad \|A\| = \lambda_{\max}(A) = \max_{0 \neq x \in \mathcal{R}^m} \frac{x^T A x}{x^T x}.$$

Using the inequalities (2.21), the extremal eigenvalues $\lambda_{\min}(A)$ and $\lambda_{\max}(A)$ satisfy

$$\alpha_{\min}(A) \left(\frac{\|x\|_X}{\|x\|} \right)^2 \leq \lambda_{\min}(A) \leq \lambda_{\max}(A) \leq \alpha_{\max}(A) \left(\frac{\|x\|_X}{\|x\|} \right)^2$$

for all vectors $x \in \mathcal{R}^m$. Since the vector norms in the finite-dimensional space $\mathcal{R}^m$ are equivalent, the corresponding equivalence constants can be used to give lower and upper bounds for the term $\frac{\|x\|_X}{\|x\|}$.

As the singular values of B are the square roots of the eigenvalues of $B^T B$, similar considerations can be made also in estimates of the extremal singular values of B.

Theorem 2.3 (min-max theorem for singular values, [42]) *If $B \in \mathcal{R}^{m,n}$, then it follows for the extremal singular values of B that*

$$\sigma_{\min}(B) = \min_{0 \neq y \in \mathcal{R}^n} \frac{\|By\|}{\|y\|}, \quad \|B\| = \sigma_{\max}(B) = \max_{0 \neq y \in \mathcal{R}^n} \frac{\|By\|}{\|y\|}.$$

Using the inequalities (2.22), we get the bounds for the extremal singular values $\sigma_{\min}(B)$ and $\sigma_{\max}(B)$ in the form

$$\beta_{\min}(B) \frac{\|y\|_Y}{\|y\|} \frac{\|By\|}{\|By\|_{X^\#}} \leq \sigma_{\min}(B) \leq \sigma_{\max}(B) \leq \beta_{\max}(B) \frac{\|y\|_Y}{\|y\|} \frac{\|By\|}{\|By\|_{X^\#}}$$

for all vectors $y \in \mathcal{R}^n$. Again, the equivalence constants can be used to give lower and upper bounds for the terms $\frac{\|y\|_Y}{\|y\|}$ and $\frac{\|By\|}{\|By\|_{X^\#}}$, respectively.

2.3   Kuhn-Karush-Tucker Systems in Interior-Point Methods

We consider the convex quadratic programming problem with equality and inequality constraints to find the vector $x \in \mathcal{R}^m$ so that

$$\min_{x \in \mathcal{R}^m} \left(\frac{1}{2} x^T A x - c^T x \right) \text{ subject to } B^T x = d, \; x \geq 0, \tag{2.23}$$

where $A \in \mathcal{R}^{m,m}$ is symmetric positive semi-definite and $B \in \mathcal{R}^{m,n}$ has a full-column rank, $c \in \mathcal{R}^m$ and $d \in \mathcal{R}^n$.

The usual transformation in interior-point methods consists in replacing the inequality constraints $x \geq 0$ with the logarithmic barriers into the form

$$\min_{x \in \mathcal{R}^m} \left(\frac{1}{2} x^T A x - c^T x - \lambda \sum_{i=1}^m \ln x_i \right), \qquad (2.24)$$

where $\lambda \geq 0$ is a barrier parameter. Having denoted the entries of matrices A and B as $A = (a_{i,j})$ and $B = (b_{j,k})$ for $i, j = 1, \ldots, m$ and $k = 1, \ldots, n$, and the entries of vectors c and d as $c = (c_1, \ldots, c_m)^T$ and $d = (d_1, \ldots, d_n)^T$, the Lagrangian function associated with (2.24) is given as

$$\mathcal{L}(x, y, \lambda) = \frac{1}{2} x^T A x - c^T x + y^T (B^T x - d) - \lambda \sum_{i=1}^m \ln x_i$$

$$= \sum_{i=1}^m x_i \left(\frac{1}{2} \sum_{j=1}^m a_{i,j} x_j - c_i \right) + \sum_{k=1}^n y_k \left(\sum_{j=1}^m B_{j,k} x_j - d_k \right)$$

$$- \lambda \sum_{i=1}^m \ln x_i,$$

where $x = (x_1, \ldots, x_m)^T$ and $y = (y_1, \ldots, y_n)^T$ is the vector of Lagrange multipliers. The first-order optimality conditions can be derived in the form

$$\frac{\partial \mathcal{L}}{\partial x_j}(x, y, \lambda) = e_j^T (Ax - c + By) - \lambda x_j^{-1} = 0, \quad j = 1, \ldots, m,$$

$$\frac{\partial \mathcal{L}}{\partial y_k}(x, y, \lambda) = e_k^T (B^T x - d) = 0, \quad k = 1, \ldots, n,$$

where $e_j \in \mathcal{R}^m$ and $e_k \in \mathcal{R}^n$ denote the j-th and k-th column of identity matrices of corresponding dimensions, respectively. Introducing the vector $z \in \mathcal{R}^m$ as $z = (z_1, \ldots, z_m)^T = \lambda (x_1^{-1}, \ldots, x_m^{-1})^T$, and the matrices $Z \in \mathcal{R}^{m,m}$ and $X \in \mathcal{R}^{m,m}$ as $Z = \text{diag}(z_1, \ldots, z_m)$ and $X = \text{diag}(x_1, \ldots, y_m)$, respectively, we can derive the optimality conditions as the system of nonlinear equations

$$Ax + By - z = c,$$

$$B^T x = d, \qquad (2.25)$$

$$XZe = \lambda e,$$

where $x \geq 0$, $z \geq 0$, and $e \in \mathcal{R}^n$ is the vector of all ones $e = (1, \ldots, 1)^T$. To solve the system of nonlinear equations (2.25), the interior-point algorithm applies the Newton method taking the initial guess x_0, y_0, and z_0 and computing the iterates x_{k+1}, y_{k+1}, and z_{k+1} for $k = 0, 1, \ldots$ in the form

$$\begin{pmatrix} x_{k+1} \\ y_{k+1} \\ z_{k+1} \end{pmatrix} = \begin{pmatrix} x_k \\ y_k \\ z_k \end{pmatrix} + \begin{pmatrix} \Delta x_k \\ \Delta y_k \\ \Delta z_k \end{pmatrix}, \tag{2.26}$$

where at each iteration, it is necessary to solve a linear system of the form

$$\begin{pmatrix} A & B & -I \\ B^T & 0 & 0 \\ Z(z_k) & 0 & X(x_k) \end{pmatrix} \begin{pmatrix} \Delta x_k \\ \Delta y_k \\ \Delta z_k \end{pmatrix} = \begin{pmatrix} c - Ax_k - By_k + z_k \\ d - B^T x_k \\ \lambda e - X(x_k)Z(z_k)e \end{pmatrix}. \tag{2.27}$$

By elimination of $\Delta z_k = (X(x_k))^{-1}[\lambda e - X(x_k)Z(z_k)e - Z(z_k)\Delta x_k]$ from the third equation of (2.27), we get the symmetric indefinite system of linear equations for unknowns Δx_k and Δy_k

$$\begin{pmatrix} A + (X(x_k))^{-1}Z(z_k) & B \\ B^T & 0 \end{pmatrix} \begin{pmatrix} \Delta x_k \\ \Delta y_k \end{pmatrix} = \begin{pmatrix} c_k \\ d_k, \end{pmatrix} \tag{2.28}$$

where $c_k = c - Ax_k - By_k + z_k + (X(x_k))^{-1}\lambda e - Z(z_k)e$ and $d_k = d - B^T x_k$. It is easy to see that (2.28) is a standard formulation of saddle-point problem in the form (1.1). For the comprehensive survey of interior-point methods including their algorithmic details, we refer to Chapter 16 in [62].

These considerations can be extended also to the convex nonlinear programming problem

$$\min_{x \in \mathcal{R}^m} f(x) \quad \text{subject to} \quad g(x) \leq 0, \tag{2.29}$$

where $f : \mathcal{R}^m \rightarrow \mathcal{R}$ and $g : \mathcal{R}^m \rightarrow \mathcal{R}^n$ with $g(x) = (g_1(x), \ldots, g_n(x))^T$ are convex and twice differentiable functions. We introduce the notation for $x = (x_1, \ldots, x_m)^T \in \mathcal{R}^m$ and for the nonnegative slack variable $z = (z_1, \ldots, z_n)^T \in \mathcal{R}^n$, and using the barrier parameter $\lambda \geq 0$, we write the inequality constraints $g(x) \leq 0$ as

$$\min_{x \in \mathcal{R}^m} \left(f(x) - \lambda \sum_{i=1}^{n} \ln z_i \right) \quad \text{subject to} \quad g(x) + z = 0. \tag{2.30}$$

The Lagrangian function associated with the problem (2.30) is given as

$$\mathcal{L}(x, y, z, \lambda) = f(x) + y^T(g(x) + z) - \lambda \sum_{i=1}^{n} \ln z_i,$$

where $y \in \mathcal{R}^n$ are the Lagrange multipliers $y = (y_1, \ldots, y_n)^T$. The stationary points for the Lagrangian function satisfy the identities

$$\frac{\partial \mathcal{L}}{\partial x_j}(x, y, z, \lambda) = \frac{\partial f(x)}{\partial x_j} + \left(\frac{\partial g_1(x)}{\partial x_j}, \ldots, \frac{\partial g_n(x)}{\partial x_j} \right) y = 0, \ j = 1, \ldots, m,$$

$$\frac{\partial \mathcal{L}}{\partial y_k}(x, y, z, \lambda) = e_k^T(g(x) + z) = 0, \ k = 1, \ldots, n,$$

$$\frac{\partial \mathcal{L}}{\partial z_k}(x, y, z, \lambda) = e_k^T(y - \lambda(Z(z))^{-1}e) = 0, \ k = 1, \ldots, n,$$

where $e_k \in \mathcal{R}^n$ denotes the k-th column of identity matrix of dimension n and where the matrix $Z(z) \in \mathcal{R}^{n,n}$ is defined $Z(z) = \text{diag}(z_1, \ldots, z_n)$. Introducing the diagonal matrix $Y(y) \in \mathcal{R}^{n,n}$ as $Y(y) = \text{diag}(y_1, \ldots, y_n)$, the first-order optimality conditions for the barrier problem represent the system of nonlinear equations

$$\nabla f(x) + (\nabla g(x))^T y = 0,$$
$$g(x) + z = 0, \tag{2.31}$$
$$Y(y)Z(z)e - \lambda e = 0,$$

where $y \geq 0$, $z \geq 0$, and $e \in \mathcal{R}^n$ denotes the vector of all ones $e = (1, \ldots, 1)^T$. The term $\nabla f(x)$ denotes the gradient $\nabla f(x) \in \mathcal{R}^m$ defined as $\nabla f(x) = \left(\frac{\partial f(x)}{\partial x_1}, \ldots, \frac{\partial f(x)}{\partial x_m} \right)^T$, and the term $\nabla g(x)$ denotes here the matrix $\nabla g(x) \in \mathcal{R}^{n,m}$ defined as $\nabla g(x) = (\nabla g_1(x), \ldots, \nabla g_n(x))^T$. The system of nonlinear equations (2.31) is solved using the Newton method with the recurrences (2.26), where at each iteration step $k = 0, 1, \ldots$, the correction vectors $(\Delta x_k, \Delta y_k, \Delta z_k)$ satisfy the linear system

$$\begin{pmatrix} A(x_k, y_k) & B(x_k) & 0 \\ B(x_k)^T & 0 & I \\ 0 & Z(z_k) & Y(y_k) \end{pmatrix} \begin{pmatrix} \Delta x_k \\ \Delta y_k \\ \Delta z_k \end{pmatrix} = - \begin{pmatrix} \nabla f(x_k) + (\nabla g(x_k))^T y_k \\ g(x_k) + z_k \\ -\lambda e + Y(y_k)Z(z_k)e \end{pmatrix}. \tag{2.32}$$

The matrix $A(x, y) \in \mathcal{R}^{m,m}$ is at some points $x = (x_1, \ldots, x_m)^T \in \mathcal{R}^m$ and $y = (y_1, \ldots, y_n)^T \in \mathcal{R}^n$ evaluated as $A(x, y) = \nabla^2 f(x) + \sum_{i=1}^{n} y_i \nabla^2 g_i(x)$, and the matrix $B(x) \in \mathcal{R}^{m,n}$ is evaluated as $B(x) = (\nabla g(x))^T$, where $\nabla^2 f(x) \in \mathcal{R}^{m,m}$ denotes the Hessian matrix defined as $\nabla^2 f(x) = \left(\frac{\partial^2 f(x)}{\partial x_i \partial x_j} \right)$ for $i, j = 1, \ldots, m$.

Again this system can be reduced eliminating the unknown $\Delta z_k = -(Y(y_k))^{-1}\lambda e +$ $Z(z_k)e - (Y(y_k))^{-1}Z(z_k)\Delta y_k$. The resulting system

$$\begin{pmatrix} A(x_k, y_k) & B(x_k) \\ B(x_k)^T & -(Y(y_k))^{-1}Z(z_k) \end{pmatrix} \begin{pmatrix} \Delta x_k \\ \Delta y_k \end{pmatrix} = -\begin{pmatrix} \nabla f(x_k) + (\nabla g(x_k))^T y_k \\ g(x_k) + (Y(y_k))^{-1}\lambda e \end{pmatrix}$$

(2.33)

has then the general 2-by-2 block structure of the form (1.2). For other details we refer to a nice presentation of methods for solution of constrained optimization problems given in the paper [13].

Chapter 3
Properties of Saddle-Point Matrices

This chapter is devoted to the study of basic algebraic properties of saddle-point matrices such as their invertibility, the existence of their block factorizations, the expressions for their inverses and to the analysis of their spectral properties such as inertia and eigenvalue localization.

3.1 Inverse of a Saddle-Point Matrix

In the following we give the necessary and sufficient condition for a saddle-point matrix to be nonsingular.

Proposition 3.1 ([12], Theorem 3.2) *Assume that A is symmetric positive semi-definite and B has a full-column rank. Then the necessary and sufficient condition for the saddle-point matrix $\mathbb{A}$ of the form (1.3) to be nonsingular is $\mathcal{N}(A) \cap \mathcal{N}(B^T) = \{0\}$, where $\mathcal{N}(A) = \{x \in \mathcal{R}^n \mid Ax = 0\}$ denotes the null-space of the matrix A and $\mathcal{N}(B^T) = \{x \in \mathcal{R}^n \mid B^T x = 0\}$ denotes the null-space of the matrix B^T.*

Proof $\Leftarrow$ Let $\mathbb{x} = \begin{pmatrix} x \\ y \end{pmatrix}$ be such that $\mathbb{A}\mathbb{x} = 0$. Hence, $Ax + By = 0$ and $B^T x = 0$, i.e., $x \in \mathcal{N}(B^T)$. It follows that

$$x^T A x = -x^T B y = -(B^T x)^T y = 0.$$

We prove that since A is symmetric positive semi-definite, $x^T A x = 0$ implies $Ax = 0$. Following Observation 7.1.6. in [42], p. 431, we consider the polynomial

$$p(\lambda) = (\lambda x + Ax)^T A(\lambda x + Ax) = \lambda^2 x^T A x + 2\lambda x^T A^2 x + x^T A^3 x$$

$$= 2\lambda \|Ax\|^2 + (Ax)^T A(Ax) \geq 0$$

M. Rozložník, *Saddle-Point Problems and Their Iterative Solution*,
Nečas Center Series, https://doi.org/10.1007/978-3-030-01431-5_3

for arbitrary real scalar λ. However, if $\|Ax\| \neq 0$, then for sufficiently large negative values of λ, we would have $p(\lambda) < 0$. We conclude that $\|Ax\| = 0$, and so $Ax = 0$. This gives $x \in \mathcal{N}(A) \cap \mathcal{N}(B^T) = \{0\}$, implying $x = 0$. Also we have $y = 0$, since $By = -Ax = 0$ and B has a full-column rank. Therefore, $\mathbb{x} = 0$ and $\mathbb{A}$ is nonsingular.

$\Rightarrow$ Assume now that $\mathcal{N}(A) \cap \mathcal{N}(B^T) \neq \{0\}$. Taking $x \neq 0$ such that $B^T x = 0$ and $Ax = 0$, we have that $\mathbb{A} \begin{pmatrix} x \\ 0 \end{pmatrix} = \begin{pmatrix} 0 \\ 0 \end{pmatrix}$. This, however, implies that $\mathbb{A}$ is singular. $\square$

Example 3.1 The saddle-point matrix $\mathbb{A} = \begin{pmatrix} 1 & 0 & 0 \\ 0 & 0 & 1 \\ 0 & 1 & 0 \end{pmatrix}$ is nonsingular with the symmetric positive semi-definite $A = \begin{pmatrix} 1 & 0 \\ 0 & 0 \end{pmatrix}$ and the off-diagonal block $B = \begin{pmatrix} 0 \\ 1 \end{pmatrix}$ satisfying $\mathcal{N}(A) = \text{span}\left\{ \begin{pmatrix} 0 \\ 1 \end{pmatrix} \right\}$, $\mathcal{N}(B^T) = \text{span}\left\{ \begin{pmatrix} 1 \\ 0 \end{pmatrix} \right\}$, and $\mathcal{N}(A) \cap \mathcal{N}(B^T) = \text{span}\left\{ \begin{pmatrix} 0 \\ 0 \end{pmatrix} \right\}$.

Example 3.2 The saddle-point matrix $\mathbb{A} = \begin{pmatrix} 1 & 0 & 1 \\ 0 & 0 & 0 \\ 1 & 0 & 0 \end{pmatrix}$ is singular with the symmetric positive semi-definite $A = \begin{pmatrix} 1 & 0 \\ 0 & 0 \end{pmatrix}$, the off-diagonal block $B = \begin{pmatrix} 1 \\ 0 \end{pmatrix}$, and $\mathcal{N}(A) \cap \mathcal{N}(B^T) = \text{span}\left\{ \begin{pmatrix} 0 \\ 1 \end{pmatrix} \right\}$.

If A is nonsingular, the 2-by-2 block matrix (1.2) can be factorized in the following three block triangular factorizations:

$$\mathbb{A} = \begin{pmatrix} A & B \\ B^T & C \end{pmatrix} = \begin{pmatrix} A & 0 \\ B^T & C - B^T A^{-1} B \end{pmatrix} \begin{pmatrix} I & A^{-1} B \\ 0 & I \end{pmatrix} \tag{3.1}$$

$$= \begin{pmatrix} I & 0 \\ B^T A^{-1} & I \end{pmatrix} \begin{pmatrix} A & B \\ 0 & C - B^T A^{-1} B \end{pmatrix} \tag{3.2}$$

$$= \begin{pmatrix} I & 0 \\ B^T A^{-1} & I \end{pmatrix} \begin{pmatrix} A & 0 \\ 0 & C - B^T A^{-1} B \end{pmatrix} \begin{pmatrix} I & A^{-1} B \\ 0 & I \end{pmatrix}. \tag{3.3}$$

Indeed, it is clear that the matrix $\mathbb{A}$ is nonsingular if and only if the Schur complement matrix $\mathbb{A} \backslash A = C - B^T A^{-1} B$ is nonsingular. Then, we have the explicit expression for the the inverse of $\mathbb{A}$ in the form

$$
\begin{pmatrix} A & B \\ B^T & C \end{pmatrix}^{-1} = \begin{pmatrix} I & A^{-1}B \\ 0 & I \end{pmatrix}^{-1} \begin{pmatrix} A^{-1} & 0 \\ 0 & (\mathbb{A}\backslash A)^{-1} \end{pmatrix} \begin{pmatrix} I & 0 \\ B^T A^{-1} & I \end{pmatrix}^{-1}
$$

$$
= \begin{pmatrix} A^{-1} + A^{-1}B(\mathbb{A}\backslash A)^{-1}B^T A^{-1} & -A^{-1}B(\mathbb{A}\backslash A)^{-1} \\ -(\mathbb{A}\backslash A)^{-1}B^T A^{-1} & (\mathbb{A}\backslash A)^{-1} \end{pmatrix}. \quad (3.4)
$$

If we assume that A is symmetric positive definite, B has a full-column rank, and C is symmetric negative semi-definite, then the Schur complement matrix $\mathbb{A}\backslash A$ is symmetric negative definite.

Using the previous identity for $C = 0$, we have also the explicit expression for the solution of the saddle-point problem (1.1) with (1.3) and (1.4):

$$
\begin{pmatrix} x_* \\ y_* \end{pmatrix} = \begin{pmatrix} A & B \\ B^T & 0 \end{pmatrix}^{-1} \begin{pmatrix} c \\ 0 \end{pmatrix} = \begin{pmatrix} \left(I + A^{-1}B(\mathbb{A}\backslash A)^{-1}B^T\right)A^{-1}c \\ -(\mathbb{A}\backslash A)^{-1}B^T A^{-1}c \end{pmatrix},
$$

where the Schur complement matrix $\mathbb{A}\backslash A = -B^T A^{-1}B$ is symmetric negative definite. Indeed, the unknown vector y_* is the solution of the symmetric negative definite system

$$
-B^T A^{-1}By = -B^T A^{-1}c \quad (3.5)
$$

that can be multiplied by -1 and converted into symmetric positive system (2.11). From $B^T x_* = 0$, the solution vector x_* belongs to the null-space $\mathcal{N}(B^T)$, i.e., it is orthogonal to $\mathcal{R}(B)$ with $x_* \perp \mathcal{R}(B)$. Given the vector y_*, we get the vector x_* as $x_* = A^{-1}(c - By_*)$, and we have the condition $B^T x_* = (A^{-1}B)^T(c - By_*) = 0$. Therefore, the vector $c - By_*$ is orthogonal to $\mathcal{R}(A^{-1}B)$ with $(c - By_*) \perp \mathcal{R}(A^{-1}B)$ or, equivalently, $(c - By_*) \perp_{A^{-1}} \mathcal{R}(B)$. Consequently, the vector y_* can be seen as the solution of the generalized least squares problem (2.10).

If we assume that A is only symmetric positive semi-definite and B has a full-column rank, then the condition $\mathcal{N}(A) \cap \mathcal{N}(B^T) = \{0\}$ implies that $\mathbb{A}$ is nonsingular. It also follows from $\mathcal{N}(A) \cap \mathcal{N}(B^T) = \{0\}$ that the matrix A is symmetric positive definite on the space $\mathcal{N}(B^T)$, i.e., $x^T Ax > 0$ for all nonzero vectors $x \in \mathcal{N}(B^T)$, $x \neq 0$. This, of course, requires $m > n$. Denote by $Z \in \mathcal{R}^{m,m-n}$ any matrix whose columns form an orthonormal basis of the null-space $\mathcal{N}(B^T)$. The projected matrix $Z^T AZ$ is thus nonsingular and symmetric positive definite. Then, the inverse of $\mathbb{A}$ can be expressed as

$$
\mathbb{A}^{-1} = \begin{pmatrix} A & B \\ B^T & 0 \end{pmatrix}^{-1} =
$$

$$
\begin{pmatrix} V & (I - VA)B(B^T B)^{-1} \\ (B^T B)^{-1}B^T(I - AV) & (B^T B)^{-1}B^T(A - AVA)B(B^T B)^{-1} \end{pmatrix},
$$

where $V = Z(Z^T A Z)^{-1} Z^T$. Note that $(B \ Z) \in \mathcal{R}^{m,m}$ is a nonsingular matrix, and due to $B^T Z = 0$ the orthogonal projector on $\mathcal{R}(B)$ can be written as $P_{\mathcal{R}(B)} = B(B^T B)^{-1} B^T = I - Z Z^T = I - P_{\mathcal{N}(B^T)}$. Then, the solution of the saddle-point problem (1.1) with (1.3) and (1.4) can be also written as

$$\begin{pmatrix} x_* \\ y_* \end{pmatrix} = \begin{pmatrix} A & B \\ B^T & 0 \end{pmatrix}^{-1} \begin{pmatrix} c \\ 0 \end{pmatrix} = \begin{pmatrix} Vc \\ (B^T B)^{-1} B^T (I - AV) c \end{pmatrix}.$$

Indeed, the unknown vector $x_* \in \mathcal{N}(B^T)$ can be expressed in the basis Z as $x_* = Z\tilde{x}_*$, where $\tilde{x}_* \in \mathcal{R}^{m-n}$. The vector of coordinates $\tilde{x}_*$ is thus the solution of the symmetric positive definite system

$$Z^T A Z \tilde{x} = Z^T c. \tag{3.6}$$

Given the vector x_*, the unknown vector y_* can be obtained as the solution of the normal equations system $B^T B y_* = B^T (c - A x_*)$. Therefore, the vector $c - A x_* - B y_*$ belongs to the null-space $\mathcal{N}(B^T)$, and it is orthogonal to $\mathcal{R}(B)$ with $(c - A x_* - B y_*) \perp \mathcal{R}(B)$. Equivalently, the vector y_* solves the least squares problem

$$\|(c - A x_*) - B y_*\| = \min_{y \in \mathcal{R}^n} \|(c - A x_*) - B y\|. \tag{3.7}$$

Another possible approach to solve the saddle-point problem (2.20) or to express the inverse of the saddle-point matrix $\mathbb{A}$ with an ill-conditioned or even singular block A is the augmented Lagrangian method. The idea is to consider the equivalent system $\mathbb{A}(W)\mathbb{x} = \mathbb{b}(W)$ defined as

$$\begin{pmatrix} A + B W B^T & B \\ B^T & 0 \end{pmatrix} \begin{pmatrix} x \\ y \end{pmatrix} = \begin{pmatrix} c + B W d \\ d \end{pmatrix}, \tag{3.8}$$

where $W \in \mathcal{R}^{n,n}$ is an appropriately chosen scaling matrix. The matrix W is usually assumed symmetric positive semi-definite so that the matrix block $A + B W B^T$ in (3.8) is nonsingular.

Proposition 3.2 ([35], Proposition 2.1) *Assume $A \in \mathcal{R}^{m,m}$ is a general square matrix of order m and B has a full-column rank such that $\mathbb{A}$ is nonsingular. Then for any nonzero scaling matrix W such that $\mathbb{A}(W)$ is nonsingular we have that*

$$\mathbb{A}^{-1}(W) = \mathbb{A}^{-1} - \begin{pmatrix} 0 & 0 \\ 0 & W \end{pmatrix} \tag{3.9}$$

with the condition number $\kappa(\mathbb{A}(W))$ satisfying the upper bound

$$\kappa(\mathbb{A}(W)) \leq \kappa(\mathbb{A}) + \|W\| \|B\|^2 \left(\|\mathbb{A}^{-1}\| + \|W\| \right) + \|\mathbb{A}\| \|W\|,$$

where $\|\cdot\|$ denotes the spectral matrix norm induced by the Euclidean vector norm.

Proof From the definition of $\mathbb{A}(W)$ we have that

$$\mathbb{A}(W) = \mathbb{A} + \begin{pmatrix} BWB^T & 0 \\ 0 & 0 \end{pmatrix}. \tag{3.10}$$

It follows from (3.9) and (3.10) that $\mathbb{A}^{-1}(W)\mathbb{A}(W) = I$. It is also clear from the triangle inequality that

$$\|\mathbb{A}(W)\| \leq \|\mathbb{A}\| + \|W\| \|B\|^2, \quad \|\mathbb{A}^{-1}(W)\| \leq \|\mathbb{A}^{-1}\| + \|W\|. \qquad \square$$

The simplest choice for the matrix W is to take a positive multiple of the identity $W = \eta I$ with $\eta > 0$ and to consider the solution of the system

$$\begin{pmatrix} A + \eta BB^T & B \\ B^T & 0 \end{pmatrix} \begin{pmatrix} x \\ y \end{pmatrix} = \begin{pmatrix} c + \eta Bd \\ d \end{pmatrix}. \tag{3.11}$$

In particular, the choice $\eta = \|A\|/\|B\|^2$ was found to work well in the sense that the condition numbers of both the blocks $A + \eta BB^T$ and $\mathbb{A}(\eta I)$ are approximately minimized. For details see [35] and [26].

3.2 Spectral Properties of Saddle-Point Matrices

If A is symmetric positive definite and B has a full-column rank, then from (3.3) we have

$$\begin{pmatrix} I & 0 \\ -B^T A^{-1} & I \end{pmatrix} \begin{pmatrix} A & B \\ B^T & 0 \end{pmatrix} \begin{pmatrix} I & -A^{-1}B \\ 0 & I \end{pmatrix} = \begin{pmatrix} A & 0 \\ 0 & -B^T A^{-1}B \end{pmatrix}. \tag{3.12}$$

Since $\mathbb{A} \in \mathcal{R}^{m+n,m+n}$ is congruent to the block diagonal matrix (3.12), it follows from that $\mathbb{A}$ is indefinite with m positive eigenvalues and n negative eigenvalues (see [42]).

As also noted in [12], this result remains true when A is only symmetric positive semi-definite, provided that the condition $\mathcal{N}(A) \cap \mathcal{N}(B^T) = \{0\}$ holds. The proof is, however, more complicated. Let

$$B = (Q \ Z) \begin{pmatrix} R \\ 0 \end{pmatrix}, \quad (Q \ Z)^T (Q \ Z) = I \tag{3.13}$$

be the QR factorization of the matrix block $B \in \mathcal{R}^{m,n}$, where $Q \in \mathcal{R}^{m,n}$ is a matrix whose columns form an orthonormal basis of the range $\mathcal{R}(B)$, and $Z \in \mathcal{R}^{m,m-n}$ is any matrix whose columns form an orthonormal basis of the null-space $\mathcal{N}(B^T)$ so that $B^T Z = 0$ and the matrix $(Z \; Q)$ is orthogonal. The matrix $\mathbb{A}$ is then congruent to the 3-by-3 block matrix

$$\begin{pmatrix} (Z \; Q)^T & 0 \\ 0 & R^{-T} \end{pmatrix} \begin{pmatrix} A & B \\ B^T & 0 \end{pmatrix} \begin{pmatrix} (Z \; Q) & 0 \\ 0 & R^{-1} \end{pmatrix}$$

$$= \begin{pmatrix} (Z \; Q)^T A (Z \; Q) & (Z \; Q)^T B \\ B^T (Z \; Q) & 0 \end{pmatrix} = \begin{pmatrix} Z^T A Z & Z^T A Q & 0 \\ Q^T A Z & Q^T A Q & I \\ 0 & I & 0 \end{pmatrix}. \qquad (3.14)$$

Due to $\mathcal{N}(A) \cap \mathcal{N}(B^T) = \{0\}$, the upper-left block $Z^T A Z \in \mathcal{R}^{m-n,m-n}$ in (3.14) is symmetric positive definite with $m - n$ positive eigenvalues. The inverse of the lower-right 2-by-2 block matrix in (3.14) is given as

$$\begin{pmatrix} Q^T A Q & I \\ I & 0 \end{pmatrix}^{-1} = \begin{pmatrix} 0 & I \\ I & -Q^T A Q \end{pmatrix}.$$

The Schur complement matrix of this matrix with respect to the matrix (3.14) is equal to $Z^T A Z$ as one can show that

$$Z^T A Z - \begin{pmatrix} Z^T A Q & 0 \end{pmatrix} \begin{pmatrix} Q^T A Q & I \\ I & 0 \end{pmatrix}^{-1} \begin{pmatrix} Q^T A Z \\ 0 \end{pmatrix} = Z^T A Z.$$

Consequently, the spectrum of the matrix $\mathbb{A}$ is given as a union of the spectra of matrices $Z^T A Z$ and $\begin{pmatrix} Q^T A Q & I \\ I & 0 \end{pmatrix}$. We will show below that the latter matrix is indefinite with n positive eigenvalues and n negative eigenvalues. In summary, the matrix $\mathbb{A}$ has m positive eigenvalues and n negative eigenvalues.

In the following we will review main results on the inclusion sets for eigenvalues of saddle-point matrices. First we consider the special case of saddle-point matrix (1.3), where the off-diagonal block B is equal to the square identity matrix of order m.

Proposition 3.3 *Let $A \in \mathcal{R}^{m,m}$ be a symmetric positive semi-definite matrix with the eigenvalues $0 \leq \lambda_m(A) \leq \cdots \leq \lambda_1(A)$, and let the matrix $B \in \mathcal{R}^{m,m}$ be equal to $B = I$. Then the eigenvalues of the matrix $\mathbb{A} = \begin{pmatrix} A & I \\ I & 0 \end{pmatrix}$ are all nonzero and they are given as*

$$\tfrac{1}{2} \left(\lambda_j(A) \pm \sqrt{\lambda_j^2(A) + 4} \right) \quad for \; j = 1, \ldots, m.$$

Proof Any eigenvalue λ of the lower-right block of the matrix (3.14) satisfies the identities

$$\begin{pmatrix} A & I \\ I & 0 \end{pmatrix} \begin{pmatrix} u \\ v \end{pmatrix} = \lambda \begin{pmatrix} u \\ v \end{pmatrix}. \tag{3.15}$$

Substituting for $u = \lambda v$ and taking into account that $\lambda \neq 0$, we obtain $Av = (\lambda - 1/\lambda)v$. Consequently, for any nonnegative eigenvalue $\lambda_j(A) \geq 0$ of the matrix A, we have the pair of one positive and one negative eigenvalue of the form $\frac{1}{2}\left(\lambda_j(A) \pm \sqrt{\lambda_j^2(A) + 4}\right)$, $j = 1, \ldots, m$. We refer to [18]. □

Further, we consider the special case of (1.3) with $A = \eta I$, where $\eta > 0$ is a positive scaling parameter. The following proposition characterizes the influence of the parameter η on the eigenvalues of the matrix $\mathbb{A}$.

Proposition 3.4 ([29], Lemma 2.1) *Let $A \in \mathcal{R}^{m,m}$ be equal to $A = \eta I$, where $\eta > 0$, and assume that $B \in \mathcal{R}^{m,n}$ has the rank$(B) = n - r$ with $n - r$ nonzero singular values $0 < \sigma_{n-r}(B) \leq \cdots \leq \sigma_1(B)$. Then the spectrum of the matrix $\mathbb{A} = \begin{pmatrix} \eta I & B \\ B^T & 0 \end{pmatrix}$ is given by*

1. *zero with multiplicity r,*
2. *η with multiplicity $m - n + r$,*
3. *$\frac{1}{2}\left(\eta \pm \sqrt{\eta^2 + 4\sigma_k^2(B)}\right)$ for $k = 1, \ldots, n - r$.*

Proof Any real eigenvalue λ of $\mathbb{A}$ must satisfy the relations $\eta u + Bv = \lambda u$ and $B^T u = \lambda v$, where $(u^T, v^T)^T$ is the corresponding (nonzero) eigenvector. Then we can distinguish three cases: If $\lambda = 0$ then $u = -\frac{1}{\eta}Bv$. Premultiplying $B^T u = \lambda v$ by v^T from the left, we get $-v^T B^T \frac{1}{\eta} Bv = 0$. Hence, $Bv = 0$, and so $u = 0$. Indeed, there are r nontrivial vectors which satisfy $Bv_k = 0$, and so there exist r zero eigenvalues with the eigenvectors of the form $(0, v_k^T)^T$. If $\lambda = \eta$, then we have $u^T BB^T u = \eta u^T Bv = 0$. Hence, $v = B^T u = 0$. This is satisfied by $m - n + r$ independent vectors u_i such that $B^T u_i = 0$ forming the eigenvectors $(u_i^T, 0)^T$. If $\lambda \neq \eta$, then we have $u = \frac{1}{\lambda - \eta} Bv$ so that $B^T Bv = \lambda(\lambda - \eta)v$. Since the matrix $B^T B$ has eigenvalues equal to the squares of the singular values of the matrix B, all real roots of the quadratic equation $\lambda(\lambda - \eta) = \sigma_k(B)^2$ for $k = 1, \ldots, n - r$ give real eigenvalues of $\mathbb{A}$. These are the remaining $2(n - r)$ eigenvalues of $\mathbb{A}$. □

Let A be nonsingular with eigenvalues $0 < \lambda_m(A) \leq \cdots \leq \lambda_1(A)$. Then it is clear that for all vectors $x \in \mathcal{R}^m$, we have

$$\lambda_m(A)\|x\|^2 \leq x^T Ax \leq \lambda_1(A)\|x\|^2.$$

Let B be of full-column rank with singular values $0 < \sigma_n(B) \leq \cdots \leq \sigma_1(B)$. Then we have for all $y \in \mathcal{R}^n$ and for all $x \in \mathcal{R}(B) = (\mathcal{N}(B^T))^\perp$

$$\sigma_n(B)\|y\| \leq \|By\| \leq \sigma_1(B)\|y\|, \quad \sigma_n(B)\|x\| \leq \|B^T x\| \leq \sigma_1(B)\|x\|.$$

The following result from Rusten and Winther [73] establishes the bounds for eigenvalues of the saddle-point matrix $\mathbb{A}$ in terms of the eigenvalues of the matrix A and in terms of the singular values of the matrix B.

Proposition 3.5 ([73], Lemma 2.1) *Let $A \in \mathcal{R}^{m,m}$ be symmetric positive definite with eigenvalues $0 < \lambda_m(A) \leq \cdots \leq \lambda_1(A)$, and let $B \in \mathcal{R}^{m,n}$ be of full-column rank with singular values $0 < \sigma_n(B) \leq \cdots \leq \sigma_1(B)$. Then the spectrum of the saddle-point matrix $\mathbb{A}$ satisfies*

$$sp(\mathbb{A}) \subset$$

$$\left[\frac{1}{2}\left(\lambda_m(A) - \sqrt{\lambda_m^2(A) + 4\sigma_1^2(B)} \right), \frac{1}{2}\left(\lambda_1(A) - \sqrt{\lambda_1^2(A) + 4\sigma_n^2(B)} \right) \right]$$

$$\cup \left[\lambda_m(A), \frac{1}{2}\left(\lambda_1(A) + \sqrt{\lambda_1^2(A) + 4\sigma_1^2(B)} \right) \right]. \tag{3.16}$$

Proof For each eigenvalue $\lambda \in \sigma(\mathbb{A})$ and its corresponding eigenvector $\begin{pmatrix} u \\ v \end{pmatrix} \neq 0$, we have

$$Au + Bv = \lambda u, \tag{3.17}$$

$$B^T u = \lambda v. \tag{3.18}$$

It is easy to see that if $u = 0$, then $v = 0$ due to $\mathrm{rank}(B) = n$ and $Bv = \lambda u - Au$. So we assume that $u \neq 0$.

Let $\lambda > 0$ be a positive eigenvalue. Then taking the inner product of (3.17) with u and the inner product of (3.18) with v, we get $u^T Au + u^T Bv = \lambda u^T u$ and $v^T B^T u = \lambda v^T v$ which leads to $u^T Au + \lambda \|v\|^2 = \lambda \|u\|^2$. Consequently we have $\lambda_m(A)\|u\|^2 + \lambda\|v\|^2 \leq \lambda\|u\|^2$ and $(\lambda_m(A) - \lambda)\|u\|^2 \leq -\lambda\|v\|^2 \leq 0$. This gives the inequality $\lambda \geq \lambda_m(A)$.

Taking the inner product of (3.18) with $B^T u$ and substituting for $u^T Bv = \frac{1}{\lambda}\|B^T u\|^2$ into $u^T Au + u^T Bv = \lambda u^T u$, we get $\lambda^2 \|u\|^2 - \lambda u^T Au - \|B^T u\|^2 = 0$. The left-hand side of previous identity can be bounded from below to get the inequality $(\lambda^2 - \lambda_1(A)\lambda - \sigma_1^2(B))\|u\|^2 \leq \lambda^2\|u\|^2 - \lambda u^T Au - \|B^T u\|^2 = 0$. This gives the bound

$$\lambda \leq \frac{1}{2}\left(\lambda_1(A) + \sqrt{\lambda_1^2(A) + 4\sigma_1^2(B)} \right).$$

Now we consider a negative eigenvalue $\lambda < 0$ and consider the inequality $(\lambda^2 - \lambda_m(A)\lambda - \sigma_1^2(B))\|u\|^2 \leq \lambda^2\|u\|^2 - \lambda u^T A u - \|B^T u\|^2 = 0$. Indeed, we have

$$\lambda \geq \frac{1}{2}\left(\lambda_m(A) - \sqrt{\lambda_m^2(A) + 4\sigma_1^2(B)}\right).$$

Finally, we take the orthogonal decomposition of $u = u|_{\mathcal{R}(B)} + u|_{\mathcal{N}(B^T)}$, where $u|_{\mathcal{R}(B)} \in \mathcal{R}(B)$ and $u|_{\mathcal{N}(B^T)} \in \mathcal{N}(B^T)$. It follows from (3.18) that $v = \frac{1}{\lambda}B^T u|_{\mathcal{R}(B)}$. Taking the inner product of (3.17) with $u|_{\mathcal{R}(B)}$ and substituting for v, we get $u|_{\mathcal{R}(B)}^T A u|_{\mathcal{N}(B^T)} = -u|_{\mathcal{R}(B)}^T A u|_{\mathcal{R}(B)} - \frac{1}{\lambda}\|B^T u|_{\mathcal{R}(B)}\|^2 + \lambda\|u|_{\mathcal{R}(B)}\|^2$. Bounding this term from below, we come to the inequality $u|_{\mathcal{R}(B)}^T A u|_{\mathcal{N}(B^T)} \geq (\lambda - \lambda_1(A) - \sigma_n^2(B)/\lambda)\|u|_{\mathcal{R}(B)}\|^2$. Taking the inner product of (3.17) with $u|_{\mathcal{N}(B^T)}$, we obtain the upper bound for $u|_{\mathcal{R}(B)}^T A u|_{\mathcal{N}(B^T)} \leq (\lambda - \lambda_m(A))\|u|_{\mathcal{N}(B^T)}\|^2 \leq 0$, and so $(\lambda^2 - \lambda_1(A)\lambda - \sigma_n^2(B))\|u|_{\mathcal{R}(B)}\|^2 \geq 0$. If $u|_{\mathcal{R}(B)} = 0$, then $v = 0$ and $Au|_{\mathcal{N}(B^T)} = \lambda u|_{\mathcal{N}(B^T)}$ with $\lambda < 0$ which cannot be the case. Therefore, we have $\lambda^2 - \lambda_1(A)\lambda - \sigma_n^2(B) \geq 0$ and

$$\lambda \leq \frac{1}{2}\left(\lambda_1(A) - \sqrt{\lambda_1^2(A) + 4\sigma_m^2(B)}\right). \qquad \square$$

Example 3.3 The 3-by-3 saddle-point matrix $\mathbb{A} = \begin{pmatrix} 1 & 0 & 0 \\ 0 & 1 & 1 \\ 0 & 1 & 0 \end{pmatrix}$ with $\lambda_1(A) = \lambda_m(A) = 1$ and $\sigma_1(B) = \sigma_n(B) = 1$ has three distinct eigenvalues 1 and $\frac{1}{2}(1 \pm \sqrt{5})$.

If $\sigma_1(B) \gg \lambda_m(A)$, then $\frac{1}{2}\left(\lambda_m(A) - \sqrt{\lambda_m^2(A) + 4\sigma_1^2(B)}\right) \approx -\sigma_1(B)$. If $\lambda_1(A) \gg \sigma_n(B)$, then $\frac{1}{2}\left(\lambda_1(A) - \sqrt{\lambda_1^2(A) + 4\sigma_n^2(B)}\right) \approx -\sigma_n(B)$. Let $\kappa(A)$ and $\kappa(B)$ be the spectral condition numbers of A and B, respectively, i.e. $\kappa(A) = \lambda_1(A)/\lambda_m(A)$ and $\kappa(B) = \sigma_1(B)/\sigma_n(B)$. The inclusion set for the eigenvalues can be scaled as follows:

$$\text{sp}(\mathbb{A}) \subset \lambda_m(A)\left[\frac{1}{2}\left(1 - \sqrt{1 + 4\varrho^2\kappa^2(B)}\right), \frac{1}{2}\left(\kappa(A) - \sqrt{\kappa^2(A) + 4\varrho^2}\right)\right]$$

$$\cup \lambda_m(A)\left[1, \frac{1}{2}\left(\kappa(A) + \sqrt{\kappa^2(A) + 4\varrho^2\kappa^2(B)}\right)\right],$$

where $\varrho = \sigma_n(B)/\lambda_m(A)$ is the ratio between the smallest singular value of B and the smallest eigenvalue of A.

In the following, we will treat two extremal cases $\lambda_m(A) = 0$ or $\sigma_n(B) = 0$.

Proposition 3.6 ([65], Proposition 1) *Let $A \in \mathcal{R}^{m,m}$ be symmetric positive semi-definite with eigenvalues $0 = \lambda_m(A) \leq \cdots \leq \lambda_1(A)$ satisfying also $x^T A x \geq \lambda_0(A)\|x\|^2$ for some $\lambda_0(A) > 0$ and for all $x \in \mathcal{N}(B^T)$. Let $B \in \mathcal{R}^{m,n}$ be of a full-column rank with singular values $0 < \sigma_n(B) \leq \cdots \leq \sigma_1(B)$. Then the spectrum of the saddle-point matrix $\mathbb{A}$ satisfies*

$$sp(\mathbb{A}) \subset \left[\frac{1}{2}\left(\lambda_m(A) - \sqrt{\lambda_m^2(A) + 4\sigma_1^2(B)} \right), \frac{1}{2}\left(\lambda_1(A) - \sqrt{\lambda_1^2(A) + 4\sigma_n^2(B)} \right) \right],$$

$$\cup \left[\lambda_0(A), \frac{1}{2}\left(\lambda_1(A) + \sqrt{\lambda_1^2(A) + 4\sigma_1^2(B)} \right) \right].$$

In the case of rank-deficient B with $\sigma_n(B) = 0$, it is clear that the standard saddle-point matrix (1.3) is singular. Therefore, one has to consider the problem with the generalized saddle-point matrix $\mathbb{A}$ in the form (1.2) introducing some symmetric negative semi-definite block $C \in \mathcal{R}^{n,n}$. Sometimes this process is called a regularization of the original saddle-point problem (1.1).

Proposition 3.7 ([76], Lemma 2.2) *Let $A \in \mathcal{R}^{m,m}$ be symmetric positive definite with eigenvalues $0 < \lambda_m(A) \leq \cdots \leq \lambda_1(A)$ and $B \in \mathcal{R}^{m,n}$ be rank-deficient with singular values $0 = \sigma_n(B) \leq \cdots \leq \sigma_1(B)$. Let $C \in \mathcal{R}^{n,n}$ be symmetric negative semi-definite with $\lambda_1(C) \leq \cdots \leq \lambda_n(C) \leq 0$ such that the matrix $B^T B - C \in \mathcal{R}^{n,n}$ is symmetric positive definite with the smallest eigenvalue $\lambda_n(B^T B - C) > 0$. Then the spectrum of the matrix $\mathbb{A} = \begin{pmatrix} A & B \\ B^T & C \end{pmatrix}$ satisfies*

$$sp(\mathbb{A}) \subset \left[\frac{1}{2}\left(\lambda_1(C) + \lambda_m(A) - \sqrt{(-\lambda_1(C) + \lambda_m(A))^2 + 4\sigma_1^2(B)} \right), \right.$$

$$\left. \frac{1}{2}\left(\lambda_1(A) - \sqrt{\lambda_1^2(A) + 4\lambda_n^2(-B^T B + C)} \right) \right],$$

$$\cup \left[\lambda_m(A), \frac{1}{2}\left(\lambda_1(A) + \sqrt{\lambda_1^2(A) + 4\sigma_1^2(B)} \right) \right].$$

Indeed, if the matrix B is rank-deficient, it makes sense to consider the solution of the saddle-point system (2.20) regularized with C in the form $C = -\eta I$ with $\eta > 0$

$$\begin{pmatrix} A & B \\ B^T & -\eta I \end{pmatrix} \begin{pmatrix} x(\eta) \\ y(\eta) \end{pmatrix} = \begin{pmatrix} c \\ d \end{pmatrix}, \tag{3.19}$$

where $x(\eta) \to x_*$ and $y(\eta) \to y_*$ as $\eta \to 0$. For details we refer to [76] and to the book [25].

The results on the spectrum of saddle-point matrices, where A is indefinite, can be found in [36]. Note that if $\mathbb{A}$ in (1.3) is nonsingular, the rank of A must be at least $m - n$ due to the condition $\mathcal{N}(A) \cap \mathcal{N}(B^T) = \{0\}$. The case when A is singular and so-called maximally rank deficient with $\text{rank}(A) = n - m$ is studied in [27].

Chapter 4
Solution Approaches for Saddle-Point Problems

Although their distinction can be rather minor in some cases, the solution algorithms for the saddle-point problems (1.1) with (1.3) can be divided into two basic subcategories called coupled (all-at-once) methods and segregated methods:

1. Coupled methods

   - do not explicitly use the block structure of the problem (1.1) and consider it as a general square system of order $m + n$;
   - deliver both components of the exact solution $\mathbb{x}_*$ or its approximate solution $\mathbb{x}_k$ for $k = 0, 1, \dots$ at once;
   - can be either direct methods using some factorization of the matrix $\mathbb{A}$ or iterative methods applied to the whole system (1.1) very often with some preconditioning technique constructed from $\mathbb{A}$.

2. Segregated methods

   - using the block elimination based on the structure of (1.1), reduce the problem to a smaller problem of dimension (n or $m - n$), and compute a component of the exact solution (either x_* or y_*) or the approximate solution (either x_k or y_k for $k = 0, 1, \dots$) to the reduced problem;
   - perform the back-substitution into the original system (1.1) to obtain the remaining component of the exact solution or its approximate solution;
   - can be either direct methods or iterative methods applied for both the reduced system and back-substitution, or their combinations when one of these two steps is computed directly and the other iteratively.

© Springer Nature Switzerland AG 2018

M. Rozložník, *Saddle-Point Problems and Their Iterative Solution*,
Nečas Center Series, https://doi.org/10.1007/978-3-030-01431-5_4

In the following we will review two main solution techniques for saddle-point systems that use a segregated approach. For details we refer to corresponding sections of [12] or to the papers [45] and [46].

4.1 Schur Complement Method

We already know from (3.3) that, if A is nonsingular and B has a full-column rank, then the saddle-point matrix $\mathbb{A}$ in (1.3) is nonsingular and the Schur complement matrix $\mathbb{A}\backslash A = -B^T A^{-1} B$ is also nonsingular. We can thus consider the block LU factorization of the saddle-point matrix $\mathbb{A}$, and multiplying (2.20) by the inverse of the block lower triangular factor in (3.2), we can perform the block elimination of the unknown x to get the transformed system

$$\begin{pmatrix} I & 0 \\ -B^T A^{-1} & I \end{pmatrix} \begin{pmatrix} A & B \\ B^T & 0 \end{pmatrix} \begin{pmatrix} x \\ y \end{pmatrix} = \begin{pmatrix} I & 0 \\ B^T A^{-1} & -I \end{pmatrix} \begin{pmatrix} c \\ d \end{pmatrix} \tag{4.1}$$

$$\Downarrow$$

$$\begin{pmatrix} A & B \\ 0 & -B^T A^{-1} B \end{pmatrix} \begin{pmatrix} x \\ y \end{pmatrix} = \begin{pmatrix} c \\ d - B^T A^{-1} c \end{pmatrix}. \tag{4.2}$$

Solving the block upper triangular system (4.2) by the block back-substitution leads to the following two main steps:

1. Solve the reduced system with the Schur complement matrix of order n in the form

$$- B^T A^{-1} B y = d - B^T A^{-1} c \tag{4.3}$$

   to get the approximate solution that is sufficiently close to its exact solution y_*.
2. Once the solution y_* has been computed from the Schur complement system (4.3), the unknown vector x_* is obtained from the solution of the system with the matrix A

$$Ax = c - By_*. \tag{4.4}$$

The systems (4.3) and (4.4) can be solved either directly or iteratively. Note that this approach is efficient if the inverse of A can be explicitly given or factorized or if the systems with A can be solved in a cheap way. In the important case of A symmetric positive definite and B of full-column rank, the negative of the Schur complement matrix $-(\mathbb{A}\backslash A) = B^T A^{-1} B$ is also symmetric positive definite.

In the case of the iterative solution of the Schur complement system (4.3), one takes an initial guess y_0 and computes the approximate solutions y_{k+1} for $k =$

$0, 1 \ldots$. If the solution of systems with the matrix A is assumed to be cheap, then in parallel we can compute also the approximate solutions x_{k+1} as the solutions of the systems $Ax_{k+1} = c - By_{k+1}$. This can be done either directly or iteratively. The Schur complement method then can be seen as a coupled method generating both the approximate solutions x_{k+1} and y_{k+1} satisfying the first block system of equations

$$Ax_{k+1} + By_{k+1} = c. \tag{4.5}$$

The residual vector $\mathbb{r}_{k+1} = \mathbb{b} - \mathbb{A}\mathbb{x}_{k+1}$ of the approximate solution $\mathbb{x}_{k+1} = \begin{pmatrix} x_{k+1} \\ y_{k+1} \end{pmatrix}$ has a particular structure with the first block component equal to zero vector, i.e., it follows from (4.5) that

$$\mathbb{r}_{k+1} = \begin{pmatrix} c \\ d \end{pmatrix} - \begin{pmatrix} A & B \\ B^T & 0 \end{pmatrix} \begin{pmatrix} x_{k+1} \\ y_{k+1} \end{pmatrix} = \begin{pmatrix} c - Ax_{k+1} - By_{k+1} \\ d - B^T x_{k+1} \end{pmatrix} = \begin{pmatrix} 0 \\ r_{k+1} \end{pmatrix}. \tag{4.6}$$

Taking into account (4.5), the vector r_{k+1} defined in (4.6) can be written as

$$r_{k+1} = d - B^T x_{k+1} = d - B^T A^{-1} c + B^T A^{-1} By_{k+1}. \tag{4.7}$$

Indeed, it is clear from (4.7) that the second block component r_{k+1} of the residual vector $\mathbb{r}_{k+1}$ in the saddle-point system (2.20) is nothing else but the residual vector of the approximate solution y_{k+1} in the Schur complement system (4.3).

We consider the standard iterative scheme, where the approximate solution y_{k+1} is computed as an update of previous approximate solution y_k with the step size α_k and the direction vector q_k in the form

$$y_{k+1} = y_k + \alpha_k q_k. \tag{4.8}$$

Substituting (4.8) into (4.7), we obtain the recurrence for updating the residual vectors r_{k+1} in the form

$$r_{k+1} = r_k + \alpha_k B^T A^{-1} Bq_k.$$

Introducing the vector p_k that satisfies the relation $Ap_k + Bq_k = 0$, we get

$$r_{k+1} = r_k - \alpha_k B^T p_k. \tag{4.9}$$

We can summarize the general inner-outer iteration Schur complement method as follows:

Algorithm 4.1 Schur complement method

---

choose y_0, solve $Ax_0 = c - By_0$

for $k = 0, 1, \ldots$

compute α_k and q_k

$y_{k+1} = y_k + \alpha_k q_k$

$\quad$ solve $Ap_k = -Bq_k$

$\quad$ back-substitution:

$\quad$ solve $Ax_{k+1} = c - By_{k+1}$

$\left.\begin{array}{l}\end{array}\right\}$ inner iteration

$r_{k+1} = r_k - \alpha_k B^T p_k$

end

outer iteration

The Schur complement method arises in various applications, and it is also called static condensation, nodal analysis, displacement method, or range-space method. For detailed discussion we refer to [12]. The details related to the implementation of the Schur complement method will be discussed in Chap. 8 on numerical behavior of saddle-point solvers.

4.2 Null-Space Method

We consider some particular solution $\hat{x} \in \mathcal{R}^m$ of the underdetermined rectangular system $B^T x = d$, where $m > n$, and denote by $Z \in \mathcal{R}^{m,m-n}$ any matrix whose columns form an orthonormal basis of the null-space $\mathcal{N}(B^T)$. The solution set of the system $B^T x = d$ can be then written as

$$x = \hat{x} + Z\tilde{x}, \tag{4.10}$$

where $\tilde{x} \in \mathcal{R}^{m-n}$ represents the coordinates of the vector $x - \hat{x}$ in the chosen basis of $\mathcal{N}(B^T)$. Substituting (4.10) into the saddle-point system (2.20), we get the saddle-point system with a particular right-hand side

$$\begin{pmatrix} A & B \\ B^T & 0 \end{pmatrix} \begin{pmatrix} Z\tilde{x} \\ y \end{pmatrix} = \begin{pmatrix} c - A\hat{x} \\ 0 \end{pmatrix}. \tag{4.11}$$

Since from (4.11) we have now $B^T Z\tilde{x} = 0$ for any vector $\tilde{x}$, we can decompose the first block equations of the saddle-point problem (4.11) into their components in the null-space $\mathcal{N}(B^T)$ and in the range $\mathcal{R}(B)$, respectively, to get the transformed system

$$\begin{pmatrix} Z^T \\ B^T \end{pmatrix} (A\ B) \begin{pmatrix} Z\tilde{x} \\ y \end{pmatrix} = \begin{pmatrix} Z^T(c - A\hat{x}) \\ B^T(c - A\hat{x}) \end{pmatrix} \tag{4.12}$$

$$\Downarrow$$

$$\begin{pmatrix} Z^T A Z & 0 \\ B^T A Z & B^T B \end{pmatrix} \begin{pmatrix} \tilde{x} \\ y \end{pmatrix} = \begin{pmatrix} Z^T(c - A\hat{x}) \\ B^T(c - A\hat{x}) \end{pmatrix}. \tag{4.13}$$

Solving the block lower triangular system (4.13) by block back-substitution corresponds to the following two steps:

1. Solve the projected system of the order $m - n$

$$Z^T A Z\tilde{x} = Z^T(c - A\hat{x}) \tag{4.14}$$

   to get the approximate solution that is sufficiently close to its exact solution $\tilde{x}_*$.
2. Once the solution $x_* = \hat{x} + Z\tilde{x}_*$ has been obtained from the solution of (4.14), the unknown vector y_* can be computed as a solution of the least squares problem $By \approx c - Ax_*$, i.e., y_* solves the problem of normal equations

$$B^T By = B^T(c - Ax_*) \Leftrightarrow \min_{y \in \mathcal{R}^n} \|(c - Ax_*) - By\|. \tag{4.15}$$

These two problems can be solved either directly or iteratively. We have shown that if $\mathbb{A}$ is nonsingular and B has a full-column rank, then the matrix A that is in general symmetric positive semi-definite must be positive definite on the null-space $\mathcal{N}(B^T)$. Therefore, the matrix $Z^T A Z$ is symmetric positive definite, and since B has a full-column rank, the least squares problem (4.15) has a unique solution.

In the case of the iterative solution of the projected system (4.14), one takes an initial guess x_0 and computes the approximate solutions x_{k+1} for $k = 0, 1, \dots$. The approximate solutions y_{k+1} can be then computed in parallel as a solution of the least squares problem system $By \approx c - Ax_{k+1}$. This can be again done either directly or iteratively. The null-space method can be seen then as a coupled method generating both the approximate solutions x_{k+1} and y_{k+1} satisfying the second block system of equations

$$B^T x_{k+1} = d, \quad x_{k+1} = \hat{x} + Z\tilde{x}_{k+1}, \tag{4.16}$$

for some vector of coordinates $\tilde{x}_{k+1} \in \mathcal{R}^{m-n}$. The residual vector $\mathbb{r}_{k+1} = \mathbb{b} -$ $\mathbb{A}\mathbb{x}_{k+1}$ of the approximate solution $\mathbb{x}_{k+1} = \begin{pmatrix} x_{k+1} \\ y_{k+1} \end{pmatrix}$ has a particular structure with the second block component equal to zero vector

$$\mathbb{r}_{k+1} = \begin{pmatrix} c \\ d \end{pmatrix} - \begin{pmatrix} A & B \\ B^T & 0 \end{pmatrix} \begin{pmatrix} x_{k+1} \\ y_{k+1} \end{pmatrix} = \begin{pmatrix} c - Ax_{k+1} - By_{k+1} \\ d - B^T x_{k+1} \end{pmatrix} = \begin{pmatrix} r_{k+1} \\ 0 \end{pmatrix}. \tag{4.17}$$

We again consider iterative scheme, where the approximate solution x_{k+1} is computed as an update of previous approximate solution x_k with the direction vector p_k in the form

$$x_{k+1} = x_k + \alpha_k p_k. \tag{4.18}$$

Substituting (4.18) into the residual r_{k+1} defined in (4.17), we get the recurrence for the vector of its coordinates $\tilde{r}_{k+1} = Z^T r_{k+1}$ in the basis Z

$$\tilde{r}_{k+1} = Z^T (c - Ax_k) - \alpha_k Z^T Ap_k = \tilde{r}_k - \alpha_k Z^T Ap_k. \tag{4.19}$$

Indeed from (4.16) the direction vector p_k belongs to $\mathcal{N}(B^T)$, and therefore, we have also the recurrence for

$$\tilde{x}_{k+1} = \tilde{x}_k + \alpha_k \tilde{p}_k, \quad \tilde{p}_k = Z^T p_k \tag{4.20}$$

and thus the vector $\tilde{r}_{k+1}$ is nothing else but the residual vector of the approximate solution $\tilde{x}_{k+1}$ in the projected system (4.14). The approximate solution y_{k+1} can be computed as a solution of the least squares problem

$$\|(c - Ax_{k+1}) - By_{k+1}\| = \min_{y \in \mathcal{R}^n} \|(c - Ax_{k+1}) - By\|. \tag{4.21}$$

It follows from (4.17) and (4.21) that $B^T r_{k+1} = B^T (c - Ax_{k+1} - By_{k+1}) = 0$. Therefore, r_{k+1} belongs to $\mathcal{N}(B^T)$, and it can be written using its vector of coordinates as $r_{k+1} = Z\tilde{r}_{k+1}$.

We can then summarize the general inner-outer iteration null-space method as follows:

Algorithm 4.2 Null-space method

---

choose x_0, solve $\min\limits_{y \in \mathcal{R}^n} \|(c - Ax_0) - By\|$

for $k = 0, 1, \ldots$

compute α_k and $p_k \in N(B^T)$

$x_{k+1} = x_k + \alpha_k p_k$

   back-substitution:

   solve $y_{k+1} = \arg\min\limits_{y \in \mathcal{R}^n} \|(c - Ax_{k+1}) - By\|$ } inner iteration

$\tilde{r}_{k+1} = \tilde{r}_k - \alpha_k Z^T A p_k$

$r_{k+1} = Z\tilde{r}_{k+1}$

end

} outer iteration

The null-space method is in various applications also called the reduced Hessian method, loop analysis, or force method. For detailed discussion we refer to the survey paper [12].

Chapter 5
Direct Solution of Saddle-Point Problems

In this chapter we give a brief overview of direct techniques used for solution of saddle-point problems. They are all based on some factorization of the saddle-point matrix. Therefore, we first recall two main factorizations of general indefinite matrices that preserve the symmetry of the original system matrix. Then we discuss cases when the 2-by-2 block matrix in the form (1.2) admits an LDL^T factorization with the diagonal factor and gives its relation to the class of symmetric quasi-definite matrices. In particular, we look at the conditioning of factors in this Cholesky-like factorization in terms of the conditioning of the matrix (1.2) and its (1, 1)-block A. Then, we show that the two main direct approaches for solving the saddle-point problems via the Schur complement method or the null-space method can be seen not just a solution procedures but also as two different factorizations of the saddle-point matrix, namely, the block LDL^T factorization and the block QTQ^T factorization, respectively.

5.1 Factorization of Symmetric Indefinite Matrices

There are several ways how to perform the Gaussian elimination on a symmetric but indefinite matrix $\mathbb{A} \in \mathcal{R}^{m+n,m+n}$ in a way that a symmetry of this factorization is preserved. The most frequent factorization is the LDL^T factorization

$$\mathbb{P}^T \mathbb{A} \mathbb{P} = \mathbb{L} \mathbb{D} \mathbb{L}^T, \tag{5.1}$$

where $\mathbb{P} \in \mathcal{R}^{m+n,m+n}$ is a permutation matrix, $\mathbb{L} \in \mathcal{R}^{m+n,m+n}$ is unit lower triangular (lower triangular with unit diagonal entries), and $\mathbb{D} \in \mathcal{R}^{m+n,m+n}$ is block diagonal with diagonal blocks of dimension 1 or 2. This factorization is essentially a symmetric block form of Gaussian elimination with symmetric pivoting. There are several possible pivoting strategies including complete pivoting or partial pivoting

© Springer Nature Switzerland AG 2018
M. Rozložník, *Saddle-Point Problems and Their Iterative Solution*,
Nečas Center Series, https://doi.org/10.1007/978-3-030-01431-5_5

with the Bunch-Parlett strategy or with the Bunch-Kaufman strategy. The second factorization is

$$\mathbb{P}^T \mathbb{A} \mathbb{P} = \mathbb{L} \mathbb{T} \mathbb{L}^T, \tag{5.2}$$

where $\mathbb{P} \in \mathcal{R}^{m+n,m+n}$ is a permutation matrix, $\mathbb{L} \in \mathcal{R}^{m+n,m+n}$ is unit lower triangular, and $\mathbb{T} \in \mathcal{R}^{m+n,m+n}$ is tridiagonal. This method is called Aasen's method. Details on the solution of general symmetric indefinite systems can be found, e.g., in Chapter 11 of the book [40].

When assuming a general symmetric but indefinite $\mathbb{A}$, the pivoting is usually needed for stability reasons. Indeed, the accuracy of computed factors depends on the size of the factor $\mathbb{L}$ in the LDL^T factorization or the factor $\mathbb{T}$ in the Aasen's method. The effects of finite precision arithmetic on these algorithms are analyzed in [40].

Example 5.1 The saddle-point matrix $\mathbb{A} = \begin{pmatrix} 0 & 1 & 1 \\ 1 & 0 & 1 \\ 1 & 1 & 0 \end{pmatrix}$ with symmetric indefinite

$A = \begin{pmatrix} 0 & 1 \\ 1 & 0 \end{pmatrix}$ has a zero main diagonal, and thus selecting the pivot of size 1 from the main diagonal is impossible, and the pivot block of size 2 is needed in the LDL^T factorization

$$\begin{pmatrix} 0 & 1 & 1 \\ 1 & 0 & 1 \\ 1 & 1 & 0 \end{pmatrix} = \begin{pmatrix} 1 & 0 & 0 \\ 0 & 1 & 0 \\ 1 & 1 & 1 \end{pmatrix} \begin{pmatrix} 0 & 1 & 0 \\ 1 & 0 & 0 \\ 0 & 0 & -2 \end{pmatrix} \begin{pmatrix} 1 & 0 & 1 \\ 0 & 1 & 1 \\ 0 & 0 & 1 \end{pmatrix}.$$

Proposition 5.1 *Let $\mathbb{A}$ be the 2-by-2 block matrix in the form (1.2), where A is symmetric positive definite, B has a full-column rank, and C is symmetric negative semi-definite. Then the matrix $\mathbb{A}$ admits the LDL^T factorization $\mathbb{P}^T A \mathbb{P} = \mathbb{L} \mathbb{D} \mathbb{L}^T$, where $\mathbb{P} = I$, i.e., no pivoting is needed, and where $\mathbb{D}$ is diagonal, i.e., there is no need for pivot blocks of size 2.*

Proof We have already noted in (3.3) that if A is nonsingular, then 2-by-2 block matrix $\mathbb{A}$ can be factorized in the block triangular LDL^T factorization

$$\mathbb{A} = \begin{pmatrix} A & B \\ B^T & C \end{pmatrix} = \begin{pmatrix} I & 0 \\ B^T A^{-1} & I \end{pmatrix} \begin{pmatrix} A & 0 \\ 0 & C - B^T A^{-1} B \end{pmatrix} \begin{pmatrix} I & A^{-1} B \\ 0 & I \end{pmatrix}.$$

Since A is symmetric positive definite, it can be decomposed with a Cholesky decomposition as $A = L_{11} D_{11} L_{11}^T$, where L_{11} is unit triangular and D_{11} is diagonal with positive entries. The Schur complement matrix $C - B^T A^{-1} B$ is symmetric negative definite, and it can be decomposed as $C - B^T A^{-1} B = -L_{22} D_{22} L_{22}^T$, where L_{22} is unit triangular and D_{22} is diagonal with positive entries. Thus we have the LDL^T factorization $\mathbb{A} = \mathbb{L}\mathbb{D}\mathbb{L}^T$, or equivalently,

$$\begin{pmatrix} A & B \\ B^T & C \end{pmatrix} = \begin{pmatrix} L_{11} & 0 \\ L_{21} & L_{22} \end{pmatrix} \begin{pmatrix} D_{11} & 0 \\ 0 & -D_{22} \end{pmatrix} \begin{pmatrix} L_{11}^T & L_{21}^T \\ 0 & L_{22}^T \end{pmatrix}, \tag{5.3}$$

where $\mathbb{D} = \begin{pmatrix} D_{11} & 0 \\ 0 & -D_{22} \end{pmatrix}$ and the off-diagonal block in the triangular factor $\mathbb{L} = \begin{pmatrix} L_{11} & 0 \\ L_{21} & L_{22} \end{pmatrix}$ is given as $L_{21} = B^T L_{11}^{-T} D_{11}^{-1}$, where $L_{11}^{-T} = (L_{11}^{-1})^T$. □

Symmetric quasi-definite matrix If C is, in addition to previous assumptions, symmetric negative definite, then the 2-by-2 block matrix $\mathbb{A}$ is called symmetric quasi-definite. Then, the LDL^T factorization (5.1) such that $\mathbb{D}$ is diagonal exists for every permutation $\mathbb{P}$, i.e., there is no need for pivot blocks of size 2. For details we refer to [80]; see also the book [63].

Alternatively, we can write the Cholesky factorization of A as $A = \tilde{L}_{11}\tilde{L}_{11}^T$ and the Cholesky factorization of the negative of the Schur complement matrix $B^T A^{-1} B - C$ as $B^T A^{-1} B - C = \tilde{L}_{22}\tilde{L}_{22}^T$, where $\tilde{L}_{11} = L_{11}D_{11}^{1/2}$ and $\tilde{L}_{22} = L_{22}D_{22}^{1/2}$ are lower triangular matrices of appropriate dimensions. Then we have the factorization $\mathbb{A} = \tilde{\mathbb{L}}\Omega\tilde{\mathbb{L}}^T = \tilde{\mathbb{R}}^T\Omega\tilde{\mathbb{R}}$, or equivalently,

$$\begin{pmatrix} A & B \\ B^T & C \end{pmatrix} = \begin{pmatrix} \tilde{L}_{11} & 0 \\ \tilde{L}_{21} & \tilde{L}_{22} \end{pmatrix} \begin{pmatrix} I & 0 \\ 0 & -I \end{pmatrix} \begin{pmatrix} \tilde{L}_{11}^T & \tilde{L}_{21}^T \\ 0 & \tilde{L}_{22}^T \end{pmatrix} \tag{5.4}$$

$$= \begin{pmatrix} \tilde{R}_{11}^T & 0 \\ \tilde{R}_{12}^T & \tilde{R}_{22}^T \end{pmatrix} \begin{pmatrix} I & 0 \\ 0 & -I \end{pmatrix} \begin{pmatrix} \tilde{R}_{11} & \tilde{R}_{12} \\ 0 & \tilde{R}_{22} \end{pmatrix}, \tag{5.5}$$

where $\Omega = \begin{pmatrix} I & 0 \\ 0 & -I \end{pmatrix}$ and where $\tilde{\mathbb{R}} = \tilde{\mathbb{L}}^T = \begin{pmatrix} \tilde{R}_{11} & \tilde{R}_{12} \\ 0 & \tilde{R}_{22} \end{pmatrix} = \begin{pmatrix} \tilde{L}_{11}^T & \tilde{L}_{21}^T \\ 0 & \tilde{L}_{22}^T \end{pmatrix}$ with the off-diagonal block $\tilde{R}_{12}$ given as $\tilde{R}_{12} = \tilde{L}_{21}^T = \tilde{L}_{11}^{-1} B$.

In the following proposition, we analyze the extremal singular values of the factor $\tilde{\mathbb{R}}$ and give a bound on its condition number $\kappa(\tilde{\mathbb{R}})$.

Proposition 5.2 ([72]) *Let* $\mathbb{A}$ *be the 2-by-2 block matrix of the form* (1.2), *where* A *is symmetric positive definite,* B *has a full-column rank, and* C *is symmetric negative semi-definite. The condition number of the factor* $\tilde{\mathbb{R}}$ *from the LDL^T factorization* $\mathbb{A} = \tilde{\mathbb{R}}^T\Omega\tilde{\mathbb{R}}$ *can be bounded as*

$$\kappa^{1/2}(\mathbb{A}) \leq \kappa(\tilde{\mathbb{R}}) \leq \|\mathbb{A}\| \left(\|\mathbb{A}^{-1}\| + 2\|A^{-1}\| \right). \tag{5.6}$$

Proof It follows immediately from (5.5) that $A = \tilde{R}_{11}^T\tilde{R}_{11}$, $B = \tilde{R}_{11}^T\tilde{R}_{12}$, and $C = \tilde{R}_{12}^T\tilde{R}_{12} - \tilde{R}_{22}^T\tilde{R}_{22}$. The Schur complement matrix $\mathbb{A}\backslash A = C - B^T A^{-1} B$ is negative definite, and it can be expressed as $\mathbb{A}\backslash A = C - \tilde{R}_{12}^T\tilde{R}_{21} = -\tilde{R}_{22}^T\tilde{R}_{22}^T$. The bound on the norm of $\tilde{\mathbb{R}}^{-1}$ can be obtained considering the identities on the inverse of $\tilde{\mathbb{R}}$ and on the inverse of $\tilde{\mathbb{R}}^T \tilde{\mathbb{R}}$

$$\tilde{\mathbb{R}}^{-1} = \begin{pmatrix} \tilde{R}_{11}^{-1} & -\tilde{R}_{11}^{-1} \tilde{R}_{12} \tilde{R}_{22}^{-1} \\ 0 & \tilde{R}_{22}^{-1} \end{pmatrix} = \begin{pmatrix} \tilde{R}_{11}^{-1} & -A^{-1} B \tilde{R}_{22}^{-1} \\ 0 & \tilde{R}_{22}^{-1} \end{pmatrix}, \qquad (5.7)$$

$$(\tilde{\mathbb{R}}^T \tilde{\mathbb{R}})^{-1} = \begin{pmatrix} A^{-1} - A^{-1} B (\mathbb{A}\backslash A)^{-1} B^T A^{-1} & A^{-1} B (\mathbb{A}\backslash A)^{-1} \\ (\mathbb{A}\backslash A)^{-1} B^T A^{-1} & -(\mathbb{A}\backslash A)^{-1} \end{pmatrix}. \qquad (5.8)$$

It is clear now that from (3.4) and (5.8), we get the identity

$$\mathbb{A}^{-1} + (\tilde{\mathbb{R}}^T \tilde{\mathbb{R}})^{-1} = 2 \begin{pmatrix} A^{-1} & 0 \\ 0 & 0 \end{pmatrix}.$$

Therefore, we have the inequalities

$$\|\mathbb{A}^{-1}\| - 2\|A^{-1}\| \le \|\tilde{\mathbb{R}}^{-1}\|^2 \le \|\mathbb{A}^{-1}\| + 2\|A^{-1}\|.$$

Using $\tilde{\mathbb{R}} = \Omega^{-1} \tilde{\mathbb{R}}^{-T} \mathbb{A}$, we can bound the norm of $\tilde{\mathbb{R}}$ also from above as

$$\|\tilde{\mathbb{R}}\| \le \|\Omega^{-1}\| \|\tilde{\mathbb{R}}^{-T}\| \|\mathbb{A}\| = \|\mathbb{A}\| \|\tilde{\mathbb{R}}^{-1}\|.$$

This gives the upper bound in (5.6). The lower bound in (5.6) can be obtained by taking into account that $\mathbb{A}^{-1} = \mathbb{R}^{-1} \Omega^{-1} \mathbb{R}^{-T}$ to get $\|\mathbb{A}^{-1}\| \le \|\mathbb{R}^{-1}\| \|\Omega^{-1}\| \|\mathbb{R}^{-T}\| = \|\tilde{\mathbb{R}}^{-1}\|^2$ due to $\|\Omega\| = \|\Omega^{-1}\| = 1$. Similarly, considering $\mathbb{A} = \tilde{\mathbb{R}}^T \Omega \tilde{\mathbb{R}}$, we can bound the norm of $\tilde{\mathbb{R}}$ from below as $\|\mathbb{A}\| \le \|\tilde{\mathbb{R}}^T\| \|\Omega\| \|\tilde{\mathbb{R}}\| = \|\tilde{\mathbb{R}}\|^2$. This completes the proof. $\square$

We also see that

$$\tilde{\mathbb{R}}^T \tilde{\mathbb{R}} = \begin{pmatrix} A & B \\ B^T & C - 2\mathbb{A}\backslash A \end{pmatrix} = \mathbb{A} - 2 \begin{pmatrix} 0 & 0 \\ 0 & \mathbb{A}\backslash A \end{pmatrix} \qquad (5.9)$$

that leads to the inequality

$$\|\tilde{\mathbb{R}}\|^2 \le \|\mathbb{A}\| + 2\|\mathbb{A}\backslash A\|, \qquad (5.10)$$

whereas the norm of the Schur complement matrix $\mathbb{A}\backslash A$ can be bounded as

$$\|\mathbb{A}\backslash A\| \le \|C\| + \|B\|^2 \|A^{-1}\| \le \|\mathbb{A}\| + \|\mathbb{A}\|^2 \|A^{-1}\|.$$

Note that it follows directly from (3.4) for the norm of $(\mathbb{A}\backslash A)^{-1}$ that

$$\|(\mathbb{A}\backslash A)^{-1}\| \le \|\mathbb{A}^{-1}\|.$$

The following proposition gives the bounds for the extremal eigenvalues of the Schur complement matrix $\mathbb{A}\backslash A$ in the particular case with $C = 0$.

Proposition 5.3 *Let $A \in \mathcal{R}^{m,m}$ be symmetric positive definite with eigenvalues $0 < \lambda_m(A) \leq \cdots \leq \lambda_1(A)$, let $B \in \mathcal{R}^{m,n}$ be of full-column rank with singular values $0 < \sigma_n(B) \leq \cdots \leq \sigma_1(B)$, and let $C \in \mathcal{R}^{n,n}$ be a zero matrix with $C = 0$. Then the spectrum of the Schur complement matrix $\mathbb{A}\backslash A = -B^T A^{-1} B$ satisfies*

$$sp(\mathbb{A}\backslash A) \subset \left[-\frac{\sigma_1^2(B)}{\lambda_m(A)}, -\frac{\sigma_n^2(B)}{\lambda_1(A)} \right]. \tag{5.11}$$

The condition number of the matrix $\mathbb{A}\backslash A$ can be bounded as $\kappa(\mathbb{A}\backslash A) \leq \kappa(A)\kappa^2(B)$.

5.2 Direct Methods for Saddle-Point Problems

In this subsection we briefly discuss coupled direct methods for the solution of the saddle-point problem (1.1) with (1.3), where A is symmetric positive definite and B has a full-column rank. Considering the Schur complement method, it is quite natural to compute the Cholesky factorization of the block A in the form $A = \tilde{L}_{11}\tilde{L}_{11}^T$, where $\tilde{L}_{11}$ is its lower triangular factor with positive diagonal entries. Then the block LDL^T factorization

$$\mathbb{A} = \begin{pmatrix} A & B \\ B^T & 0 \end{pmatrix} = \begin{pmatrix} I & 0 \\ B^T A^{-1} & I \end{pmatrix} \begin{pmatrix} A & 0 \\ 0 & -B^T A^{-1} B \end{pmatrix} \begin{pmatrix} I & A^{-1}B \\ 0 & I \end{pmatrix}$$

can be written as

$$\mathbb{A} = \begin{pmatrix} I & 0 \\ B^T \tilde{L}_{11}^{-T} \tilde{L}_{11}^{-1} & I \end{pmatrix} \begin{pmatrix} \tilde{L}_{11}\tilde{L}_{11}^T & 0 \\ 0 & -B^T \tilde{L}_{11}^{-T} \tilde{L}_{11}^{-1} B \end{pmatrix} \begin{pmatrix} I & \tilde{L}_{11}^{-T} \tilde{L}_{11}^{-1} B \\ 0 & I \end{pmatrix}.$$

Given the factor $\tilde{L}_{11}$, we compute the columns of the matrix $\tilde{L}_{11}^{-1} B$ by back-substitution and form the Schur complement matrix as

$$\mathbb{A}\backslash A = -B^T \tilde{L}_{11}^{-T} \tilde{L}_{11}^{-1} B = -(\tilde{L}_{11}^{-1} B)^T (\tilde{L}_{11}^{-1} B).$$

Since its negative is also symmetric positive definite, it makes sense to compute the Cholesky factorization $-(\mathbb{A}\backslash A) = \tilde{L}_{22}\tilde{L}_{22}^T$, where $\tilde{L}_{22}$ is lower triangular with positive diagonal entries so that we have the Cholesky-like factorization of the saddle-point matrix

$$\mathbb{A} = \begin{pmatrix} \tilde{L}_{11} & 0 \\ (\tilde{L}_{11}^{-1}B)^T & \tilde{L}_{22} \end{pmatrix} \begin{pmatrix} I & 0 \\ 0 & -I \end{pmatrix} \begin{pmatrix} \tilde{L}_{11}^T & \tilde{L}_{11}^{-1}B \\ 0 & \tilde{L}_{22}^T \end{pmatrix}. \tag{5.12}$$

Substituting the factorization (5.12) into the saddle-point system (2.20) and applying the forward substitution on its right-hand side, we get the upper triangular system

$$
\begin{pmatrix} \tilde{L}_{11}^T & \tilde{L}_{11}^{-1} B \\ 0 & -\tilde{L}_{22}^T \end{pmatrix} \begin{pmatrix} x \\ y \end{pmatrix} = \begin{pmatrix} \tilde{L}_{11}^{-1} c \\ \tilde{L}_{22}^{-1} \left(d - (\tilde{L}_{11}^{-1} B)^T (\tilde{L}_{11}^{-1} c) \right) \end{pmatrix} \tag{5.13}
$$

which is solved with the back-substitution. Note that other factorizations of the block A such as the QR factorization are also possible and the corresponding Schur complement method can be derived. For a brief overview of direct methods based on triangular factorizations of the diagonal block A and the Schur complement matrix $\mathbb{A} \backslash A$, we refer to Section 7 of [12].

As it was already noted, one can apply direct methods also in the null-space method. If $m > n$, then an orthonormal basis of $\mathcal{N}(B^T)$ can be computed by means of the QR factorization of the off-diagonal block $B \in \mathcal{R}^{m,n}$ in the form

$$
B = Q \begin{pmatrix} R \\ 0 \end{pmatrix} = (Y \ Z) \begin{pmatrix} R \\ 0 \end{pmatrix} = YR, \tag{5.14}
$$

where $Q \in \mathcal{R}^{m,m}$ is orthogonal matrix, the columns of $Y \in \mathcal{R}^{m,n}$ form an orthonormal basis of $\mathcal{R}(B)$, the columns of $Z \in \mathcal{R}^{m,m-n}$ form an orthonormal basis of $\mathcal{N}(B^T)$, and $R \in \mathcal{R}^{n,n}$ is upper triangular. Substituting (5.14) into the saddle-point problem (2.20), we get

$$
\begin{pmatrix} A & Q \begin{pmatrix} R \\ 0 \end{pmatrix} \\ (R^T \ 0) Q^T & 0 \end{pmatrix} \begin{pmatrix} x \\ y \end{pmatrix} = \begin{pmatrix} c \\ d \end{pmatrix}. \tag{5.15}
$$

The saddle-point matrix (1.3) can be then factorized as

$$
\begin{aligned}
\mathbb{A} = \begin{pmatrix} A & B \\ B^T & 0 \end{pmatrix} &= \begin{pmatrix} Q & 0 \\ 0 & R^T \end{pmatrix} \begin{pmatrix} Y^T AY & Y^T AZ & I \\ Z^T AY & Z^T AZ & 0 \\ I & 0 & 0 \end{pmatrix} \begin{pmatrix} Q^T & 0 \\ 0 & R \end{pmatrix} \\
&= \begin{pmatrix} Q & 0 \\ 0 & I \end{pmatrix} \begin{pmatrix} Y^T AY & Y^T AZ & R \\ Z^T AY & Z^T AZ & 0 \\ R^T & 0 & 0 \end{pmatrix} \begin{pmatrix} Q^T & 0 \\ 0 & I \end{pmatrix}. \tag{5.16}
\end{aligned}
$$

The relation of the block QTQ^T factorization of (5.16), where $\mathbb{T}$ is block anti-triangular, to the null-space method was thoroughly reviewed in [70]; see also [66]. We can decompose the unknown vector x in the corresponding components in $\mathcal{R}(B)$ and $\mathcal{N}(B^T)$ to get

$$
\begin{pmatrix} Y^T AY & Y^T AZ & I \\ Z^T AY & Z^T AZ & 0 \\ I & 0 & 0 \end{pmatrix} \begin{pmatrix} Y^T x \\ Z^T x \\ Ry \end{pmatrix} = \begin{pmatrix} Y^T c \\ Z^T c \\ R^{-T} d \end{pmatrix}.
$$

The permutation of the first and third block equations gives the block lower triangular system of equations

$$
\begin{pmatrix} I & 0 & 0 \\ Z^T A Y & Z^T A Z & 0 \\ Y^T A Y & Y^T A Z & I \end{pmatrix} \begin{pmatrix} Y^T x \\ Z^T x \\ R y \end{pmatrix} = \begin{pmatrix} R^{-T} d \\ Z^T c \\ Y^T c \end{pmatrix}.
\tag{5.17}
$$

The block lower triangular system (5.17) is solved successively, where in the first step $Y^T x = R^{-T} d$, we find a particular solution $\hat{x} \in \mathcal{R}(B)$ to the underdetermined system $B^T x = d$. Then we solve the projected system $Z^T A Z \tilde{x} = Z^T (c - A\hat{x})$, where $\tilde{x} = Z^T x$ to get the unknown vector $x_* = \hat{x} + Z\tilde{x}_*$. This is usually solved by the Cholesky factorization of the symmetric positive definite matrix $Z^T A Z = U^T U$ and by the subsequent forward substitution and back-substitution that are equivalent to the upper triangular systems $U\tilde{x} = U^{-T} Z^T (c - A\hat{x})$. Finally, we compute the unknown vector y_* as a solution of the upper triangular system $R y = Y^T (c - A x_*)$ which is equivalent to the solution of the system of normal equations $B^T B y = B^T (c - A x_*)$.

One can also consider the LU factorization of $B \in \mathcal{R}^{m,n}$ in the form

$$
PBQ = \hat{L} \begin{pmatrix} \hat{U} \\ 0 \end{pmatrix} = \begin{pmatrix} \hat{L}_{11} & 0 \\ \hat{L}_{21} & I \end{pmatrix} \begin{pmatrix} \hat{U} \\ 0 \end{pmatrix},
\tag{5.18}
$$

where $\hat{U} \in \mathcal{R}^{n,n}$ is upper triangular and $\hat{L} \in \mathcal{R}^{m,m}$ is nonsingular lower triangular matrix (with $\hat{L}_{11} \in \mathcal{R}^{n,n}$ and $\hat{L}_{21} \in \mathcal{R}^{m-n,n}$) generated by the Gaussian elimination applied to the permuted matrix PBQ and where $P \in \mathcal{R}^{m,m}$ and $Q \in \mathcal{R}^{n,n}$ are permutation matrices that are given by the numerical pivoting and sparsity considerations. For the sake of simplicity, we set $P = I$ and $Q = I$ here, and we assume that the matrix $\mathbb{A}$ and the right-hand side vector $\mathbb{b}$ have been consistently permuted. It is easy to show that the columns of the matrix $Z = \hat{L}^{-T} \begin{pmatrix} 0 \\ I \end{pmatrix} \in \mathcal{R}^{m,m-n}$ form a basis of $\mathcal{N}(B^T)$. Similarly, the columns of the matrix $Y = \hat{L}^{-T} \begin{pmatrix} I \\ 0 \end{pmatrix} \in \mathcal{R}^{m,n}$ form a basis of $\mathcal{R}(B)$. The saddle-point matrix $\mathbb{A}$ can be then factorized as

$$
\mathbb{A} = \begin{pmatrix} \hat{L} & 0 \\ 0 & \hat{U}^T \end{pmatrix} \begin{pmatrix} \hat{L}^{-1} A \hat{L}^{-T} & \begin{pmatrix} I \\ 0 \end{pmatrix} \\ (I\ 0) & 0 \end{pmatrix} \begin{pmatrix} \hat{L}^T & 0 \\ 0 & \hat{U} \end{pmatrix}.
\tag{5.19}
$$

Substituting (5.19) into the saddle-point system (2.20) and applying the forward substitution on its right-hand side, we get

$$
\begin{pmatrix} \hat{L}^{-1} A \hat{L}^{-T} & \begin{pmatrix} I \\ 0 \end{pmatrix} \\ (I\ 0) & 0 \end{pmatrix} \begin{pmatrix} \hat{L}^T x \\ \hat{U} y \end{pmatrix} = \begin{pmatrix} \hat{L}^{-1} c \\ \hat{U}^{-T} d \end{pmatrix}.
\tag{5.20}
$$

The permutation of the first and second block equations gives the block lower triangular system of equations

$$
\begin{pmatrix} (I\ 0) & 0 \\ \hat{L}^{-1}A\hat{L}^{-T} & \begin{pmatrix} I \\ 0 \end{pmatrix} \end{pmatrix} \begin{pmatrix} \hat{L}^T x \\ \hat{U} y \end{pmatrix} = \begin{pmatrix} \hat{U}^{-T}d \\ \hat{L}^{-1}c \end{pmatrix}.
\tag{5.21}
$$

Since $\hat{L}^{-1} = \begin{pmatrix} \hat{L}_{11}^{-1} & 0 \\ -\hat{L}_{21}\hat{L}_{11}^{-1} & I \end{pmatrix}$ and $\hat{L}^{-T} = \begin{pmatrix} \hat{L}_{11}^{-T} & -\hat{L}_{11}^{-T}\hat{L}_{21}^T \\ 0 & I \end{pmatrix}$, the block lower triangular system (5.21) is solved by forward substitution, where we get all the components of $\hat{L}^T x$ and $\hat{U} y$. The unknown vectors x_* and y_* are obtained then using the corresponding back-substitutions. For details on the null-space method based on the factorization (5.14), we refer to [1]. The null-space method using the factorization (5.18) is discussed in papers [2] and [3].

The following easy-to-prove proposition gives the bounds for the extremal eigenvalues of the matrix $Z^T A Z$ in terms of the extremal eigenvalues of A and in terms of the extremal singular values of the matrix Z, where we assume that the column vectors of Z form a general non-orthogonal basis of $\mathcal{N}(B^T)$ satisfying only the condition $B^T Z = 0$.

Proposition 5.4 *Let $A \in \mathcal{R}^{m,m}$ be symmetric positive definite with eigenvalues $0 < \lambda_m(A) \leq \cdots \leq \lambda_1(A)$, and let $Z \in \mathcal{R}^{m,m-n}$, $m > n$ be of full-column rank with singular values $0 < \sigma_{m-n}(Z) \leq \cdots \leq \sigma_1(Z)$ such that $B^T Z = 0$. Then the spectrum of the matrix $Z^T A Z$ satisfies*

$$
sp(Z^T A Z) \subset \left[\lambda_m(A)\sigma_{m-n}^2(Z), \lambda_1(A)\sigma_1^2(Z) \right].
\tag{5.22}
$$

The condition number of the matrix $Z^T A Z$ can be bounded as $\kappa(Z^T A Z) \leq \kappa(A)\kappa^2(Z)$. If the column vectors of Z form an orthonormal basis of $\mathcal{N}(B^T)$, then it follows trivially that $\kappa(Z^T A Z) \leq \kappa(A)$.

Chapter 6
Iterative Solution of Saddle-Point Problems

Although sparse direct solvers are very competitive, they can be less efficient for challenging problems due to their storage and computational limitations. If we cannot solve the saddle-point problem directly, in many applications, we have to use some iterative method. Coupled iterative methods applied to the system (1.1) take some initial guess $x_0 \in \mathcal{R}^{m+n}$ and generate approximate solutions $x_k \in \mathcal{R}^{m+n}$ for $k = 1, \ldots$ such that they satisfy $x_k \to x_*$. The convergence to the exact solution $x_* \in \mathcal{R}^{m+n}$ can be also measured using the residual vectors given as $r_k = b - Ax_k$, where we eventually have $r_k \to 0$.

We will distinguish here between three important classes of iterative methods: stationary methods, Krylov subspace methods, and multigrid methods.

1. Stationary methods

   - are used as stand-alone saddle-point solvers,
   - are used as preconditioners for Krylov subspace methods,
   - are used as smoothers in multigrid methods.

2. Krylov subspace methods

   - are used most often as efficient stand-alone saddle-point solvers,
   - are used to accelerate the convergence of stationary methods,
   - need preconditioners to improve their performance and reliability.

3. Multigrid methods

   - need smoothers, typically stationary iterative methods, on fine grid problems,
   - are used as stand-alone solvers, but they need direct or Krylov subspace methods for coarse grid problems,
   - are often used as preconditioners for Krylov subspace methods, if the coarse grid problems are solved inexactly.

© Springer Nature Switzerland AG 2018
M. Rozložník, *Saddle-Point Problems and Their Iterative Solution*,
Nečas Center Series, https://doi.org/10.1007/978-3-030-01431-5_6

6.1   Stationary Iterative Methods

We consider the saddle-point problem (1.1) with (1.3) and assume that A is symmetric positive definite and B has a full-column rank. Given some initial guess $\mathbb{x}_0$, a stationary iterative method can be viewed as the fixed-point iteration scheme

$$\mathbb{x}_{k+1} = \mathbb{G}\mathbb{x}_k + \mathbb{c}, \quad \mathbb{G} = \mathbb{M}^{-1}\mathbb{N}, \quad \mathbb{c} = \mathbb{M}^{-1}\mathbb{b} \tag{6.1}$$

or, equivalently, the iterative refinement scheme

$$\mathbb{x}_{k+1} = \mathbb{x}_k + \mathbb{M}^{-1}\mathbb{r}_k, \quad \mathbb{r}_k = \mathbb{b} - \mathbb{A}\mathbb{x}_k, \tag{6.2}$$

where $\mathbb{A} = \mathbb{M} - \mathbb{N}$ is some splitting of the saddle-point matrix $\mathbb{A}$ and $k = 0, 1, \ldots$.

In the following, we discuss schemes, where the splitting matrix $\mathbb{M}$ is block lower triangular. One of the most classical and widely used schemes is the Uzawa's method. The matrices $\mathbb{M}$ and $\mathbb{N}$ are defined as

$$\mathbb{M} = \begin{pmatrix} A & 0 \\ B^T & -\frac{1}{\beta}I \end{pmatrix}, \quad \mathbb{N} = \begin{pmatrix} 0 & -B \\ 0 & -\frac{1}{\beta}I \end{pmatrix}, \tag{6.3}$$

where $\beta > 0$ is a positive relaxation parameter. The corresponding matrices $\mathbb{G}$ and $\mathbb{H}$ from (6.1) and (6.2) are given as

$$\mathbb{G} = \begin{pmatrix} 0 & -A^{-1}B \\ 0 & I - \beta B^T A^{-1}B \end{pmatrix}, \quad \mathbb{M}^{-1} = \begin{pmatrix} A^{-1} & 0 \\ \beta B^T A^{-1} & -\beta I \end{pmatrix}.$$

Considering both vector components of the initial guess x_0 and y_0, the Uzawa's method corresponds then to the coupled iteration scheme

$$\begin{pmatrix} x_{k+1} \\ y_{k+1} \end{pmatrix} = \begin{pmatrix} 0 & -A^{-1}B \\ 0 & I - \beta B^T A^{-1}B \end{pmatrix} \begin{pmatrix} x_k \\ y_k \end{pmatrix} + \begin{pmatrix} A^{-1} & 0 \\ \beta B^T A^{-1} & -\beta I \end{pmatrix} \begin{pmatrix} c \\ d \end{pmatrix}$$

$$= \begin{pmatrix} A^{-1}(c - By_k) \\ y_k - \beta \left(d - B^T A^{-1}(c - By_k) \right) \end{pmatrix}$$

$$= \begin{pmatrix} A^{-1}(c - By_k) \\ y_k - \beta(d - B^T x_{k+1}) \end{pmatrix}.$$

It is clear that the update $y_{k+1} = y_k - \beta(d - B^T A^{-1}c + B^T A^{-1}By_k)$ represents the Richardson update in the Richardson method applied to the Schur complement system (4.3), whereas

$$y_{k+1} = y_k + M^{-1}r_k, \quad M^{-1} = -\beta I, \quad r_k = d - B^T A^{-1}c + B^T A^{-1}By_k.$$

Since A is symmetric positive definite and B has a full-column rank, the matrix $B^T A^{-1} B$ is also symmetric positive definite. The matrix $I - \beta B^T A^{-1} B$ is thus symmetric, and having denoted the maximal eigenvalue and the minimal eigenvalue of $B^T A^{-1} B$ by $\lambda_1(B^T A^{-1} B)$ and $\lambda_n(B^T A^{-1} B)$, respectively, we get

$$(1 - \beta\lambda_1(B^T A^{-1} B))\|y\|^2 \le ((I - \beta B^T A^{-1} B)y, y) \le (1 - \beta\lambda_n(B^T A^{-1} B))\|y\|^2$$

for any vector $y \in \mathcal{R}^n$. Consequently, the norm of $I - \beta B^T A^{-1} B$ is equal to $\|I - \beta B^T A^{-1} B\| = \max\{|1 - \beta\lambda_1(B^T A^{-1} B)|, |1 - \beta\lambda_n(B^T A^{-1} B)|\}$.

We trivially have $1 - \beta\lambda_n(B^T A^{-1} B) < 1$. Assuming $-1 < 1 - \beta\lambda_1(B^T A^{-1} B)$, we can show that Richardson's iteration converges for all values β satisfying $0 < \beta < \frac{2}{\lambda_1(B^T A^{-1} B)}$. For $\beta = \frac{1}{\lambda_1(B^T A^{-1} B)}$, we have $\|I - \beta B^T A^{-1} B\| = 1 - \frac{1}{\kappa(B^T A^{-1} B)}$, where $\kappa(B^T A^{-1} B) = \frac{\lambda_1(B^T A^{-1} B)}{\lambda_n(B^T A^{-1} B)}$. One can also show that the spectral radius of the matrix $I - \beta B^T A^{-1} B$ is minimized for

$$\beta_* = \frac{2}{\lambda_1(B^T A^{-1} B) + \lambda_n(B^T A^{-1} B)}$$

and it is equal to

$$\|I - \beta_* B^T A^{-1} B\| = \frac{\kappa(B^T A^{-1} B) - 1}{\kappa(B^T A^{-1} B) + 1}.$$

The Uzawa's iteration method is given in the following algorithm:

Algorithm 6.1 Uzawa's method

> choose x_0 and y_0
>
> for $k = 0, 1, 2, \ldots$
>
> $\quad x_{k+1} = A^{-1}(c - By_k)$
>
> $\quad y_{k+1} = y_k - \beta(d - B^T x_{k+1})$
>
> end

The Arrow-Hurwicz method is an alternative to Uzawa's method. It is based on the splitting $\mathbb{A} = \mathbb{M} - \mathbb{N}$, where

$$\mathbb{M} = \begin{pmatrix} \frac{1}{\alpha}I & 0 \\ B^T & -\frac{1}{\beta}I \end{pmatrix}, \quad \mathbb{N} = \begin{pmatrix} \frac{1}{\alpha}I - A & -B \\ 0 & -\frac{1}{\beta}I \end{pmatrix}, \tag{6.4}$$

and where $\alpha > 0$ is also a positive relaxation parameter. The scheme (6.2) can be then written as

$$
\begin{pmatrix} x_{k+1} \\ y_{k+1} \end{pmatrix} = \begin{pmatrix} x_k \\ y_k \end{pmatrix} + \begin{pmatrix} \alpha I & 0 \\ \alpha \beta B^T & -\beta I \end{pmatrix} \begin{pmatrix} c - A x_k - B y_k \\ d - B^T x_k \end{pmatrix}
$$

$$
= \begin{pmatrix} x_k + \alpha(c - A x_k - B y_k) \\ y_k + \alpha \beta B^T (c - A x_k - B y_k) - \beta(d - B^T x_k) \end{pmatrix}
$$

$$
= \begin{pmatrix} x_k + \alpha(c - A x_k - B y_k) \\ y_k - \beta(d - B^T x_{k+1}) \end{pmatrix}.
$$

The Arrow-Hurwicz iteration method reads as follows:

Algorithm 6.2 Arrow-Hurwicz method

---

choose x_0 and y_0

for $k = 0, 1, 2, \ldots$

$\quad x_{k+1} = x_k + \alpha(c - A x_k - B y_k)$

$\quad y_{k+1} = y_k - \beta(d - B^T x_{k+1})$

end

The inexact Uzawa's method was considered and analyzed in detail by Zulehner in [85]. It can be seen as a stationary iterative method (6.1) or (6.2) with the splitting

$$
\mathbb{M} = \begin{pmatrix} \hat{A} & 0 \\ B^T & -\hat{C} \end{pmatrix}, \quad \mathbb{N} = \begin{pmatrix} \hat{A} - A & -B \\ 0 & -\hat{C} \end{pmatrix}, \tag{6.5}
$$

where $\hat{A}$ is a symmetric positive definite approximation to A such that $A - \hat{A}$ is also symmetric positive definite and $\hat{C}$ is a symmetric positive definite approximation to $B^T A^{-1} B$. This includes the exact Uzawa's method, where $\hat{A} = A$ and $\hat{C} = \frac{1}{\beta}$, and the Arrow-Hurwicz method, where $\hat{A} = \frac{1}{\alpha} I$ and $\hat{C} = \frac{1}{\beta}$. The case of inexact Uzawa-type splitting, where $\hat{C} = I$, was considered in [16] as a truly pioneering approach in preconditioning of saddle-point problems. This concept was further generalized in [85]. Under assumptions that there exists a constant $\hat{\alpha} > 0$ such that

$$
(\hat{A}x, x) < (Ax, x) \le (1 + \hat{\alpha})(\hat{A}x, x) \tag{6.6}
$$

for all $x \in \mathcal{R}^m$, $x \ne 0$, and that there exist constants $\hat{\gamma}$ and $\hat{\delta}$ satisfying $0 < \hat{\gamma} \le \hat{1} + \delta$ and

$$
\hat{\gamma}(\hat{C}y, y) \le (B^T A^{-1} B y, y) \le (1 + \hat{\delta})(\hat{C}y, y) \tag{6.7}
$$

for all $y \in \mathcal{R}^n$, it was shown in [85] that the matrix $\mathbb{G} = \mathbb{M}^{-1}\mathbb{N}$ is $\mathbb{H}$-symmetric, i.e., it is symmetric with the inner product induced by the matrix

$$\mathbb{H} = \begin{pmatrix} \hat{A} - A & 0 \\ 0 & \hat{C} \end{pmatrix} \tag{6.8}$$

satisfying the condition $\mathbb{G}^T\mathbb{H} = \mathbb{H}\mathbb{G}$. If $\hat{\delta} + \hat{\alpha}(3 + \hat{\delta}) < 1$, then the spectral radius $\varrho(\mathbb{G})$ satisfies $\varrho(\mathbb{G}) < 1$, and the scheme (6.1) is convergent for any initial guess $\mathbb{x}_0$. In addition, it was shown in [85] that the matrix $\mathbb{M}^{-1}\mathbb{A} = \mathbb{I} - \mathbb{G}$ is symmetric positive definite with respect to the inner product induced by $\mathbb{H}$, i.e., it holds

$$\left(\mathbb{M}^{-1}\mathbb{A}\mathbb{x}, \mathbb{x}\right)_{\mathbb{H}} = \left(\mathbb{M}^{-1}\mathbb{A}\mathbb{x}, \mathbb{H}\mathbb{x}\right) > 0 \tag{6.9}$$

for all vectors $\mathbb{x} \in \mathcal{R}^{m+n}$, $\mathbb{x} \neq 0$. For details we refer to Theorem 4.1 in [85]. For an overview of literature on Uzawa's method, Arrow-Hurwicz method, and related methods, we refer to Subsection 8.1 in [12].

The second class of stationary iterative methods is based on the splitting $\mathbb{A} = \mathbb{M} - \mathbb{N}$, where the splitting matrix $\mathbb{M}$ is indefinite and nonsingular. We start with the scheme related to the Schur complement method. The splitting is given as

$$\mathbb{M} = \begin{pmatrix} A & B \\ B^T & B^T A^{-1} B - E \end{pmatrix}, \quad \mathbb{N} = \begin{pmatrix} 0 & 0 \\ 0 & B^T A^{-1} B - E \end{pmatrix}, \tag{6.10}$$

where $E \in \mathcal{R}^{n,n}$ is an easily invertible matrix. In many cases one takes $E = I$ or E diagonal with nonzero diagonal entries. The main goal here is that the indefinite splitting matrix $\mathbb{M}$ and its inverse can be factorized using two block-triangular factors

$$\mathbb{M} = \begin{pmatrix} A & 0 \\ B^T & -E \end{pmatrix}\begin{pmatrix} I & A^{-1}B \\ 0 & I \end{pmatrix}, \quad \mathbb{M}^{-1} = \begin{pmatrix} I & -A^{-1}B \\ 0 & I \end{pmatrix}\begin{pmatrix} A & 0 \\ B^T & -E \end{pmatrix}^{-1}.$$

The iteration matrix $\mathbb{G}$ in the scheme (6.1) has the form

$$\mathbb{G} = \begin{pmatrix} 0 & -A^{-1}B(I - E^{-1}B^T A^{-1}B) \\ 0 & I - E^{-1}B^T A^{-1}B \end{pmatrix}$$

and we have the following proposition on a sufficient condition for the convergence of the scheme.

Proposition 6.1 ([20]) *If $\|I - E^{-1/2}B^T A^{-1}B E^{-1/2}\| < 1$, then the spectral radius of $\mathbb{G}$ is $\varrho(\mathbb{G}) < 1$, and the scheme (6.1) is convergent for any initial guess $\mathbb{x}_0$.*

The next scheme is related to the null-space method, and it is based on splitting

$$\mathbb{M} = \begin{pmatrix} D & B \\ B^T & 0 \end{pmatrix}, \quad \mathbb{N} = \begin{pmatrix} D - A & 0 \\ 0 & 0 \end{pmatrix}, \tag{6.11}$$

where $D \in \mathcal{R}^{m,m}$ is an easily invertible matrix. In many cases we have $D = I$ or D diagonal with nonzero diagonal entries. The iteration matrix $\mathbb{G}$ in the scheme (6.1) has then the form

$$\mathbb{G} = \begin{pmatrix} (D^{-1} - D^{-1}B(B^T D^{-1}B)^{-1}B^T D^{-1})(D - A) & 0 \\ (B^T D^{-1}B)^{-1}B^T D^{-1}(D - A) & 0 \end{pmatrix},$$

whereas the indefinite splitting matrix $\mathbb{M}$ and its inverse are again factorized using two block-triangular factors

$$\mathbb{M} = \begin{pmatrix} D & 0 \\ B^T & -B^T D^{-1}D \end{pmatrix} \begin{pmatrix} I & D^{-1}B \\ 0 & I \end{pmatrix},$$

$$\mathbb{M}^{-1} = \begin{pmatrix} I & -D^{-1}B \\ 0 & I \end{pmatrix} \begin{pmatrix} D & 0 \\ B^T & -B^T D^{-1}B \end{pmatrix}^{-1}.$$

We have the following proposition.

Proposition 6.2 ([15]) *If $\|I - D^{-1/2}AD^{-1/2}\| < 1$, then the spectral radius of $\mathbb{G}$ is $\varrho(\mathbb{G}) < 1$, and the scheme (6.1) is convergent for any initial guess $\mathbb{x}_0$.*

A generalization of the splitting (6.10) was discussed by Zulehner in [84] and [85] in the context of stationary iterative methods used as smoothers in the multigrid method. These methods are based on the splitting

$$\mathbb{M} = \begin{pmatrix} \hat{A} & B \\ B^T & B^T \hat{A}^{-1}B - \hat{C} \end{pmatrix} = \begin{pmatrix} \hat{A} & 0 \\ B^T & -\hat{C} \end{pmatrix} \begin{pmatrix} I & \hat{A}^{-1}B \\ 0 & I \end{pmatrix}, \tag{6.12}$$

$$\mathbb{N} = \begin{pmatrix} \hat{A} - A & -B \\ 0 & B^T \hat{A}^{-1}B - \hat{C} \end{pmatrix}, \tag{6.13}$$

where $\hat{A}$ is a symmetric positive definite approximation to A such that $A - \hat{A}$ is symmetric positive definite and $\hat{C}$ is a symmetric positive definite approximation to $B^T \hat{A}^{-1}B$ such that $B^T \hat{A}^{-1}B - \hat{C}$ is also symmetric positive definite. Note that this includes the case (6.11), where $\hat{A} = A$ and $\hat{C} = E$, and also the case (6.11), where $\hat{A} = D$ and $\hat{C} = B^T D^{-1}B$. Under assumptions that there exists a constant $0 < \hat{\beta} < 1$ such that

$$\hat{\beta}(\hat{A}x, x) \leq (Ax, x) < (\hat{A}x, x) \tag{6.14}$$

for all $x \in \mathcal{R}^m$, $x \neq 0$ and that there exists a constant $\hat{\delta} \leq 0$ satisfying

$$(\hat{C}y, y) < (B^T \hat{A}^{-1} By, y) \leq (1 + \hat{\delta})(\hat{C}y, y) \tag{6.15}$$

for all $y \in \mathcal{R}^n$, $y \neq 0$, it was shown in [85] (see also [84]) that the matrix $\mathbb{G} = \mathbb{M}^{-1}\mathbb{N}$ is $\mathbb{N}$-symmetric satisfying the condition $\mathbb{G}^T\mathbb{N} = \mathbb{N}\mathbb{G}$. If $\hat{\delta}(3 - 2\hat{\beta}) < 1$, then the spectral radius $\varrho(\mathbb{G})$ satisfies $\varrho(\mathbb{G}) = \|\mathbb{G}\|_\mathbb{N} < 1$. In addition, the matrix $\mathbb{M}^{-1}\mathbb{A} = \mathbb{I} - \mathbb{G}$ is symmetric positive definite with respect to the inner product induced by $\mathbb{N}$, i.e., for all vectors $x \in \mathcal{R}^{m+n}$, $x \neq 0$, we have

$$\left(\mathbb{M}^{-1}\mathbb{A}x, x\right)_\mathbb{N} = \left(\mathbb{M}^{-1}\mathbb{A}x, \mathbb{N}x\right) > 0. \tag{6.16}$$

For details we refer to Theorem 5.2 in [85].

Define the matrices $\mathbb{A} = \begin{pmatrix} A & B \\ B^T & 0 \end{pmatrix}$ and $\mathbb{K} = \begin{pmatrix} I & 0 \\ 0 & -I \end{pmatrix}$. The system (1.1) can be transformed into the system $(\mathbb{K}\mathbb{A})x = \mathbb{K}b$, where

$$\mathbb{K}\mathbb{A} = \begin{pmatrix} I & 0 \\ 0 & -I \end{pmatrix} \begin{pmatrix} A & B \\ B^T & 0 \end{pmatrix} = \begin{pmatrix} A & B \\ -B^T & 0 \end{pmatrix}.$$

If the matrix A is skew-symmetric, i.e., $A^T = -A$, then the matrix $\mathbb{K}\mathbb{A}$ is skew-symmetric, i.e., $(\mathbb{K}\mathbb{A})^T = -\mathbb{K}\mathbb{A}$.

Proposition 6.3 ([12], p. 24, Theorem 3.6) *Assume that A is symmetric positive semi-definite, B has a full-column rank, and $\mathcal{N}(A) \cap \mathcal{N}(B^T) = \{0\}$. Then $\mathbb{K}\mathbb{A}$ is nonsymmetric but positive semi-definite.*

Proof Observe that for any $x \in \mathcal{R}^{m+n}$, we have $x^T\mathbb{K}\mathbb{A}x = x^T \left[\frac{\mathbb{K}\mathbb{A}+(\mathbb{K}\mathbb{A})^T}{2}\right] x$ and

$$\frac{\mathbb{K}\mathbb{A} + (\mathbb{K}\mathbb{A})^T}{2} = \begin{pmatrix} A & 0 \\ 0 & 0 \end{pmatrix}. \qquad \square$$

It was shown in [11] that for any parameter η, the matrix $\mathbb{K}\mathbb{A}$ is symmetric with respect to the bilinear form defined by the matrix $\mathbb{H} = \begin{pmatrix} A - \eta I & B \\ B^T & \eta I \end{pmatrix}$, i.e., it satisfies $\mathbb{H}(\mathbb{K}\mathbb{A}) = (\mathbb{K}\mathbb{A})^T\mathbb{H}$. The extension taking the nonzero matrix C in (1.2) was derived in [53].

Next, we consider the system (1.1) with the 2-by-2 block matrix (1.2) and the transformed system $\mathbb{K}\mathbb{A}x = \mathbb{K}b$ with

$$\mathbb{K}\mathbb{A} = \begin{pmatrix} I & 0 \\ 0 & -I \end{pmatrix} \begin{pmatrix} A & B \\ B^T & C \end{pmatrix} = \begin{pmatrix} A & B \\ -B^T & -C \end{pmatrix}, \qquad \mathbb{K}b = \begin{pmatrix} c \\ -d \end{pmatrix},$$

where A is in general nonsymmetric but positive definite, B has a full-column rank, and C is symmetric negative definite. Given the initial guess x_0 and two splittings of the matrix $\mathbb{K}\mathbb{A}$ in the form $\mathbb{K}\mathbb{A} = \mathbb{M}_1 - \mathbb{N}_1 = \mathbb{M}_2 - \mathbb{N}_2$, we use the two-step iteration scheme defined by two successive recurrences

$$x_{k+1/2} = \mathbb{M}_1^{-1}\left(\mathbb{N}_1 x_k + \mathbb{K}\mathbb{b}\right), \tag{6.17}$$

$$x_{k+1} = \mathbb{M}_2^{-1}\left(\mathbb{N}_2 x_{k+1/2} + \mathbb{K}\mathbb{b}\right). \tag{6.18}$$

Indeed, we can eliminate the intermediate vector $x_{k+1/2}$ to get

$$x_{k+1/2} = \mathbb{M}_1^{-1}\mathbb{N}_1 x_k + \mathbb{M}_1^{-1}\mathbb{K}\mathbb{b} = x_k + \mathbb{M}_1^{-1}\mathbb{K}(\mathbb{b} - \mathbb{A}x_k).$$

Then the two-step scheme (6.17) and (6.18) can be equivalently written as a fixed-point iteration (6.1) in the form

$$x_{k+1} = \mathbb{G}x_k + \mathbb{c} = \mathbb{M}_2^{-1}\mathbb{N}_2\mathbb{M}_1^{-1}\mathbb{N}_1 x_k + \mathbb{M}_2^{-1}(\mathbb{N}_2\mathbb{M}_1^{-1} + \mathbb{I})\mathbb{K}\mathbb{b}.$$

The Hermitian/skew-Hermitian splitting (HSS) is actually the splitting of the matrix $\mathbb{K}\mathbb{A}$ into its symmetric and skew-symmetric part

$$\mathbb{K}\mathbb{A} = \begin{pmatrix} H(A) & 0 \\ 0 & -C \end{pmatrix} + \begin{pmatrix} S(A) & B \\ -B^T & 0 \end{pmatrix} \equiv \mathbb{H}(A) + \mathbb{S}(A). \tag{6.19}$$

Note that the matrix $\mathbb{H}(A)$ is symmetric positive definite, if the matrices $H(A) = \frac{1}{2}(A + A^T)$ and $-C$ are symmetric positive definite. It is also clear that the matrix $S(A) = \frac{1}{2}(A - A^T)$ is skew-symmetric with $(S(A))^T = -S(A)$. The HSS iteration method uses the two-step scheme (6.17) and (6.18), where we consider the splittings

$$\mathbb{K}\mathbb{A} = \mathbb{H}(A) + \omega\mathbb{I} - (\omega\mathbb{I} - \mathbb{S}(A)) = \mathbb{M}_1 - \mathbb{N}_1, \tag{6.20}$$

$$\mathbb{K}\mathbb{A} = \mathbb{S}(A) + \omega\mathbb{I} - (\omega\mathbb{I} - \mathbb{H}(A)) = \mathbb{M}_2 - \mathbb{N}_2, \tag{6.21}$$

where $\omega > 0$ is a positive parameter. Its recurrences are then given as

$$x_{k+1/2} = (\mathbb{H}(A) + \omega\mathbb{I})^{-1}\left((\omega\mathbb{I} - \mathbb{S}(A))x_k + \mathbb{K}\mathbb{b}\right), \tag{6.22}$$

$$x_{k+1} = (\mathbb{S}(A) + \omega\mathbb{I})^{-1}\left((\omega\mathbb{I} - \mathbb{H}(A))x_{k+1/2} + \mathbb{K}\mathbb{b}\right). \tag{6.23}$$

Note that the matrix $\mathbb{H}(A) + \omega\mathbb{I}$ is symmetric positive definite and the matrix $\mathbb{S}(A) + \omega\mathbb{I}$ is shifted skew-symmetric. For details we refer to [8].

It was shown in [8] that if the matrix $\mathbb{H}(A)$ is positive definite, then this stationary iteration method converges for all $\omega > 0$. This would require C to be negative definite, but it was also shown that it is enough if $H(A)$ is symmetric positive definite, B has a full-column rank, and C has negative semi-definite (possibly zero). If $\mathbb{H}(\mathbb{A})$ is positive definite, then the choice $\omega_* = \sqrt{\lambda_1(\mathbb{H}(\mathbb{A}))\lambda_{m+n}(\mathbb{H}(\mathbb{A}))}$ minimizes the upper bound for the spectral radius of the matrix $\mathbb{G}$. Its explicit formula is given as

$$\mathbb{G} = \mathbb{I} - \frac{1}{2\omega}(\mathbb{H}(\mathbb{A}) + \omega\mathbb{I})(\mathbb{S}(\mathbb{A}) + \omega\mathbb{I})\mathbb{A}.$$

6.2 Krylov Subspace Methods

We consider a system of linear equations

$$\mathcal{A}x = b, \tag{6.24}$$

where $\mathcal{A}$ is a nonsingular matrix and b a right-hand side vector. In the context of solving saddle-point problems, the system (6.24) can be the whole saddle-point system (1.1) with $\mathcal{A} = \mathbb{A}$, the Schur complement system (4.3) with $\mathcal{A} = \mathbb{A} \backslash A$, or the projected system (4.14) with $\mathcal{A} = Z^T A Z$.

Starting from an initial guess x_0 with the residual $r_0 = b - \mathcal{A}x_0$, we build the sequence of nested spaces $\mathcal{K}_k$ (approximation spaces) and $\mathcal{L}_k$ (constraint spaces) for $k = 0, 1, \ldots$ and compute the sequence of approximate solutions x_k such that they satisfy the Petrov-Galerkin condition

$$x_k \in x_0 + \mathcal{K}_k, \quad r_k = b - \mathbb{A}x_k \perp \mathcal{L}_k. \tag{6.25}$$

The iterative methods that are nowadays applied to large-scale linear systems are mostly Krylov subspace methods. This class of methods corresponds to $\mathcal{K}_k = \mathcal{K}_k(\mathcal{A}, r_0)$, where $\mathcal{K}_k(\mathcal{A}, r_0)$ denotes the k-th Krylov subspace associated with $\mathcal{A}$ and r_0 defined as $\mathcal{K}_k(\mathcal{A}, r_0) = \text{span}(r_0, \mathcal{A}r_0, \ldots, \mathcal{A}^{k-1}r_0)$. It follows then from (6.25) that for the error $x_* - x_k$ and for the residual r_k, we have $x_* - x_k = p_k(\mathcal{A})(x_* - x_0)$ and $r_k = p_k(\mathcal{A})r_0$, respectively, where p_k stands for some polynomial of degree at most k satisfying $p_k(0) = 1$. The whole class of such polynomials will be denoted as P_k.

Based on the choice of the spaces $\mathcal{L}_k$, we will distinguish between the following three subclasses of Krylov subspace methods:

- Orthogonal residual methods – $\mathcal{L}_k = \mathcal{K}_k(\mathcal{A}, r_0)$ (CG method for symmetric positive definite systems [39], FOM method for nonsymmetric systems [74]);
- Minimal residual methods – $\mathcal{L}_k = \mathcal{A}\mathcal{K}_k(\mathcal{A}, r_0)$ (MINRES method for symmetric indefinite systems [64], GMRES method for nonsymmetric systems [75]);
- Biorthogonalization methods – $\mathcal{L}_k = \mathcal{K}_k(\mathcal{A}^T, r_0)$ (Bi-CG method for nonsymmetric systems [77, 78], QMR method for nonsymmetric systems [30]).

Iterative methods for solving large linear systems are reviewed in [32]. The main focus is on developments in the area of conjugate gradient-type algorithms and iterative methods for nonsymmetric matrices. The book [37] includes the most useful iterative methods from a practical point of view and focuses on their theoretical analysis. A detailed treatment of mathematical principles behind derivation and analysis of Krylov subspace methods is given in the monograph [54]. Practical aspects and implementations of various iterative methods are discussed in [74].

Symmetric positive definite systems As we already know, the negative of the Schur complement matrix $\mathcal{A} = -(\mathbb{A}\backslash A)$ in (4.3) as well as the projected matrix $\mathcal{A} = Z^T A Z$ in (4.14) is symmetric positive definite. The most known and widely used Krylov subspace method for solving systems with symmetric positive definite matrix is the conjugate gradient (CG) method introduced by Hestenes and Stiefel in [39]. This method generates the approximate solutions (6.25) satisfying the equivalent energy norm error minimization conditions

$$r_k \perp \mathcal{K}_k(\mathcal{A}, r_0) \Leftrightarrow \|x_* - x_k\|_{\mathcal{A}} = \min_{x \in x_0 + \mathcal{K}_k(\mathcal{A}, r_0)} \|x_* - x\|_{\mathbb{A}} \tag{6.26}$$

$$\Leftrightarrow \|p_k(\mathcal{A})(x_* - x_0)\|_{\mathcal{A}} = \min_{\substack{p \in P_k \\ p(0)=1}} \|p(\mathcal{A})(x_* - x_0)\|_{\mathcal{A}}. \tag{6.27}$$

Using $x_* - x_k = p_k(\mathcal{A})(x_* - x_0)$ and taking an expansion of the initial error $x_* - x_0$ into orthogonal eigenvector basis of $\mathcal{A} = \mathcal{V}\mathcal{D}\mathcal{V}^T$ with $\mathcal{V}^T\mathcal{V} = \mathcal{V}\mathcal{V}^T = I$ and with diagonal $\mathcal{D}$ containing the eigenvalues of $\mathcal{A}$, we obtain the bound for the energy norm of the error

$$\|x_* - x_k\|_{\mathcal{A}} = \|p_k(\mathcal{A})(x_* - x_0)\|_{\mathcal{A}} = \min_{p \in P_k} \|\mathcal{V}p(\mathcal{D})\mathcal{V}^T(x_* - x_0)\|_{\mathcal{A}}$$

$$\leq \min_{p \in P_k} \|p(\mathcal{D})\| \|x_* - x_0\|_{\mathcal{A}} = \min_{p \in P_k} \max_{\lambda \in \text{sp}(\mathcal{A})} |p(\lambda)| \|x_* - x_0\|_{\mathcal{A}}. \tag{6.28}$$

The bound (6.28) is sharp in the sense that for every k there exists an initial guess x_0 (dependent on k) such that equality holds (see the pages 50–51 in the book [37]). The convergence of the CG method thus depends on the eigenvalue distribution of the matrix $\mathcal{A}$. The right-hand side of (6.28) can be further estimated considering a polynomial approximation problem on the minimal interval $[\beta, \alpha]$ that covers the spectrum $\text{sp}(\mathcal{A}) \subset [\beta, \alpha]$, where $0 < \beta \leq \alpha$. Solving this problem, we obtain the bound

$$\min_{\substack{p \in P_k \\ p(0)=1}} \max_{\lambda \in \text{sp}(\mathcal{A})} |p(\lambda)| \leq 2 \left(\frac{\sqrt{\alpha} - \sqrt{\beta}}{\sqrt{\alpha} + \sqrt{\beta}} \right)^k. \tag{6.29}$$

Taking ideally $\alpha = \lambda_{\max}(\mathcal{A})$ and $\beta = \lambda_{\min}(\mathcal{A})$, we get a well-known bound for the energy norm of the relative error in the CG method

$$\frac{\|x_* - x_k\|_{\mathcal{A}}}{\|x_* - x_0\|_{\mathcal{A}}} \leq 2 \left(\frac{\sqrt{\kappa(\mathcal{A})} - 1}{\sqrt{\kappa(\mathcal{A})} + 1} \right)^k . \tag{6.30}$$

The upper bound (6.30) is often used in applications for general description of the convergence rate of the CG method. However, it was pointed out (see the discussion in Chapter 5.6 of [54] or in Chapter 11 of [56]) that using such an asymptotic relation has principal limitations and any consideration concerning the rate of convergence relevant to practical computations must also include analysis of effects of rounding errors; see [34].

Symmetric indefinite systems   When the matrix $\mathcal{A}$ is symmetric but not positive definite as it is the case if we consider $\mathcal{A} = \mathbb{A}$, the most frequently used method is the minimal residual (MINRES) method proposed in [64]. The MINRES method computes the approximate solutions (6.25) satisfying the equivalent residual norm minimization conditions

$$r_k \perp \mathcal{A}\mathcal{K}_k(\mathcal{A}, r_0) \Leftrightarrow \|b - \mathcal{A}x_k\| = \min_{x \in x_0 + \mathcal{K}_k(\mathcal{A}, r_0)} \|b - \mathcal{A}x\| \tag{6.31}$$

$$\Leftrightarrow \|p_k(\mathcal{A})r_0\| = \min_{\substack{p \in P_k \\ p(0)=1}} \|p(\mathcal{A})r_0\|. \tag{6.32}$$

Although the CG method has been proposed for solving symmetric positive definite system, it can be applied also to symmetric but indefinite system. Actually it was shown in [64] that the approximate solution of CG in such case exists at least at every second step. In addition, there exists a relation between the CG method and MINRES method (see p. 86 in [37]). It appears that the norms of their residuals are for $k < n$ mutually connected via

$$\frac{\|r_k^{MINRES}\|}{\|r_k^{CG}\|} = \sqrt{1 - \left(\frac{\|r_k^{MINRES}\|}{\|r_{k-1}^{MINRES}\|} \right)^2} . \tag{6.33}$$

This relation describes the "peak/plateau" behavior for this pair of methods in the case of a general symmetric (indefinite) system, where the CG method may have large residual norm or even some of its approximate solutions may not exist (then $\|r_k^{CG}\|$ is considered as infinitely large and (6.33) in this sense still holds). Usually such situation is accompanied with the jump (peak) in the convergence curve of CG, and we have $\|r_k^{MINRES}\| \ll \|r_k^{CG}\|$. Then (6.33) implies the stagnation (plateau) of the residual norm in the MINRES method $\|r_k^{MINRES}\| \approx \|r_{k-1}^{MINRES}\|$. On the other hand, when MINRES converges quickly, then the CG residual is comparable to that in the MINRES method. An excellent overview of symmetric Krylov subspace methods can be found in [78].

Using analogous approach as in (6.28) for the CG method, we can obtain the bound for the residual norm in the MINRES method

$$\|r_k\| = \min_{\substack{p \in P_k \\ p(0)=1}} \|\mathcal{V} p(\mathcal{D}) \mathcal{V}^T r_0\| = \min_{\substack{p \in P_k \\ p(0)=1}} \|p(\mathcal{D}) \mathcal{V}^T r_0\|$$

$$\leq \min_{\substack{p \in P_k \\ p(0)=1}} \|p(\mathcal{D})\| \|\mathcal{V}^T r_0\| = \min_{\substack{p \in P_k \\ p(0)=1}} \max_{\lambda \in sp(\mathcal{A})} |p(\lambda)| \|r_0\|. \qquad (6.34)$$

This bound is sharp in the same way as (6.28), and so the convergence rate of MINRES depends on the eigenvalue distribution of the matrix $\mathcal{A}$. Since $\mathcal{A}$ is symmetric but indefinite, the inclusion set for the spectrum is formed by two disjoint intervals $[-\alpha, -\beta] \cup [\gamma, \delta]$, where $0 < \beta < \alpha$ and $0 < \gamma < \delta$ (one interval on the positive and one interval on the negative side of the real axis). The polynomial approximation problem on this set has always a unique solution [82], but the optimal polynomial is analytically known only in special cases such as $sp(\mathcal{A}) \subset [-\alpha, -\beta] \cup [\gamma, \delta]$, $0 < \beta \leq \alpha$, $0 < \gamma \leq \delta$, $\alpha - \beta = \delta - \gamma$. In this case, we have the upper bound

$$\min_{\substack{p \in P_k \\ p(0)=1}} \max_{\lambda \in sp(\mathcal{A})} |p(\lambda)| \leq 2 \left(\frac{\sqrt{\alpha\delta} - \sqrt{\beta\gamma}}{\sqrt{\alpha\delta} + \sqrt{\beta\gamma}} \right)^{\left[\frac{k}{2}\right]}. \qquad (6.35)$$

Therefore, it is significantly a harder task to get a reasonable practical bound for the MINRES method. Frequently one can estimate only the convergence rate using the asymptotic convergence factor [82, 83] defined as $\lim_{k \to \infty} \left(\frac{\|r_k\|}{\|r_0\|} \right)^{\frac{1}{k}}$ and satisfying the bound

$$\lim_{k \to \infty} \left(\frac{\|r_k\|}{\|r_0\|} \right)^{\frac{1}{k}} \leq \lim_{k \to \infty} \left(\min_{\substack{p \in P_k \\ p(0)=1}} \max_{\lambda \in sp(\mathcal{A})} |p(\lambda)| \right)^{\frac{1}{k}}. \qquad (6.36)$$

As it was shown in [83], such bounds are quite descriptive when estimating the convergence rate of minimum residual methods for saddle-point problem which depend on an asymptotically small parameter such as the mesh size in a finite difference or finite element discretization; see also [82] or [58].

Nonsymmetric systems In the nonsymmetric case, the situation is even less transparent. Many iterative methods have been proposed, but in practice only a few of them are really used. The GMRES method [75] is a direct generalization of the MINRES method to the nonsymmetric case. It is also based on the residual norm minimization (6.31). If we assume that the system matrix $\mathcal{A}$ is diagonalizable with $\mathcal{A} = \mathcal{V}\mathcal{D}\mathcal{V}^{-1}$ and consider the residual vector $r_k = p_k(\mathcal{A})r_0$ with the expansion of

r_0 into eigenvector basis $\mathcal{V}$, we get

$$\frac{\|r_k\|}{\|r_0\|} \leq \min_{\substack{p \in P_k \\ p(0)=1}} \|\mathcal{V} p(\mathcal{D}) \mathcal{V}^{-1}\| \leq \kappa(\mathcal{V}) \min_{\substack{p \in P_k \\ p(0)=1}} \max_{\lambda \in \mathrm{sp}(\mathcal{A})} |p(\lambda)|. \tag{6.37}$$

Therefore, if the condition number $\kappa(\mathcal{V})$ of the eigenbasis $\mathcal{V}$ is "reasonably bounded," one can use (6.37) similarly to (6.34) as in the symmetric case, where the convergence rate is determined by the eigenvalue distribution of the system matrix (see, e.g., pp. 134–135 in [74]). The principal difficulties of any GMRES convergence analysis which is based on eigenvector expansion of the initial residual when the eigenvector matrix is ill-conditioned were pointed out in the convergence analysis of GMRES when applied to tridiagonal Toeplitz systems [51]. However, in many cases, the system matrix is non-diagonalizable, and arguments about the density of the class of diagonalizable matrices in the whole matrix space are not applicable for extending the bound (6.37) to arbitrary non-diagonalizable (or nonnormal) system. Actually, one can get any convergence behavior of GMRES independently on the spectrum [38, 54]. A frequently used bound in applications is based on the field of values of the matrix $\mathcal{A}$ (see p. 195 in [74]). Provided that the field of values does not contain the origin, we have for the relative residual norm the bound

$$\frac{\|r_k\|}{\|r_0\|} \leq \left[1 - \left(\frac{\min_{x \neq 0}(\mathcal{A}x, x)}{\max_{x \neq 0}(\mathcal{A}x, x)} \right)^2 \right]^{\frac{k}{2}}. \tag{6.38}$$

The field of values of the matrix $\mathcal{A}$ can be analyzed in terms of discretization parameters, and the convergence rate of the GMRES can be estimated via (6.38) for some applications such as in [23]. The majority of approaches used to analyze the convergence of GMRES are based only on the matrix of the discretized system, and they do not take into account any influence of the right-hand side. The necessity of considering also the right-hand side was emphasized in the convergence analysis of GMRES for a model convection-diffusion problem; see [52]. For a detailed discussion of various approaches used in convergence analysis of GMRES, we refer Chapter 5.7 of [54] and references therein.

The GMRES method can be implemented only using full-term recurrences which significantly limit its practical applicability [74, 75]. Therefore, nonsymmetric iterative methods with short-term recurrences are used. They generate the approximate solutions (6.25) which are, however, not optimal in the sense of the energy error norm (6.26) or residual norm minimization (6.31). Nevertheless, although these methods may not converge and are difficult to analyze, they usually work on real-world problems. The most important methods are the biconjugate gradient (Bi-CG) method [77, 78] and the quasi-minimal residual (QMR) method [30, 32]. There is no significant difference in the behavior of these two methods. The slightly modified "peak/plateau" relation for the Bi-CG and QMR methods

$$\frac{\|s_k^{QMR}\|}{\|r_k^{Bi-CG}\|} = \sqrt{1 - \left(\frac{\|s_k^{QMR}\|}{\|s_{k-1}^{QMR}\|}\right)^2} \tag{6.39}$$

holds for $k < n$. The quantity s_k^{QMR} denotes the so-called quasi-residual in the QMR method. The detailed analysis of the relation between Bi-CG and QMR residuals can be found in [37]. The fact that there is no substantial difference in using iterative schemes is even more profound for preconditioned Krylov subspace methods, where the efficiency of the solver is actually determined not by the choice of a particular method but by the choice of a preconditioner. Clearly, an efficient preconditioner "hides" all local differences between various methods, which most often show very similar convergence behavior (ideally one observes a termination after several iteration steps).

6.3   Preconditioned Krylov Subspace Methods

Convergence of Krylov subspace methods is accelerated, and their robustness can be increased in practical applications by a suitable preconditioning. Roughly speaking, preconditioning is the transformation of $\mathcal{A}x = b$ into another system that is easier to solve. The preconditioner itself is often represented by the matrix $\mathcal{P}$ that should be a "good" approximation to $\mathcal{A}$ so that chosen Krylov subspace method has better convergence properties on the resulting preconditioned system. In addition, its inverse of $\mathcal{P}^{-1}$ should be easily computable or at least the systems with $\mathcal{P}$ should be easily solvable.

We can distinguish between the following three main preconditioning approaches: left preconditioning leading to the preconditioned system

$$\mathcal{P}^{-1}\mathcal{A}x = \mathcal{P}^{-1}b, \tag{6.40}$$

right preconditioning with the preconditioned system

$$\mathcal{A}\mathcal{P}^{-1}z = b, \quad x = \mathcal{P}^{-1}z, \tag{6.41}$$

and, provided that $\mathcal{P}$ can be factorized as $\mathcal{P} = \mathcal{P}_1\mathcal{P}_2$, two-sided preconditioning leading to the preconditioned system

$$\mathcal{P}_1^{-1}\mathcal{A}\mathcal{P}_2^{-1}y = \mathcal{P}_1^{-1}b, \quad x = \mathcal{P}_2^{-1}y. \tag{6.42}$$

The concept of preconditioning itself is not new, and many preconditioning techniques have been proposed in the last several decades. The quality of the preconditioner depends on how much information from the original problem we can use. The range of problems, that can be treated by a particular preconditioner, is therefore often limited. In general, we can distinguish between two main classes of preconditioners. Pure algebraic preconditioners are based on algebraic techniques such as incomplete factorizations, sparse approximate inverses, or algebraic multilevel approaches, while application-dependent preconditioners heavily use the information from the underlying continuous problem. A basic overview of preconditioning schemes can be found in survey papers [10] and [81] and in the corresponding chapters of books [37] and [74]. For a recent survey on preconditioning in the context of saddle-point problems, we refer to [67].

As we have pointed out in previous subsection, the convergence of iterative methods for symmetric systems depends on the distribution of eigenvalues of the system matrix. A better distribution of eigenvalues and/or reduced conditioning of the preconditioned matrix often leads to a fast convergence. For nonsymmetric systems, the information on a spectrum of preconditioned system may not be enough to ensure fast convergence. In some cases, the preconditioning techniques lead to sufficiently better field of values or to the reduction of a minimal polynomial degree of the preconditioned matrix.

Symmetric positive definite system + symmetric positive definite preconditioner It seems quite natural that in the case of the symmetric positive definite system (6.24), in our saddle-point context, if $\mathcal{A} = -(\mathbb{A} \backslash A)$ or $\mathcal{A} = Z^T A Z$, is the preconditioning matrix $\mathcal{P}$ chosen also as symmetric positive definite. Then there exists its square root $\mathcal{P}^{1/2}$ that is also symmetric, and the preconditioned system (6.42) in the two-sided preconditioning can be rewritten as

$$\mathcal{P}^{-1/2} \mathcal{A} \mathcal{P}^{-1/2} y = \mathcal{P}^{-1/2} b, \quad y = \mathcal{P}^{1/2} x. \tag{6.43}$$

The matrix $\mathcal{P}^{-1/2} \mathcal{A} \mathcal{P}^{-1/2}$ is again symmetric positive definite, and one can apply the same iterative method as in the unpreconditioned case. The natural method of choice here is the CG method [39]. The straightforward application of the CG method on (6.43) would lead to a sequence of approximate solutions y_k to the vector $y_* = \mathcal{P}^{1/2} x_*$, but we want to compute the solution $x_* = \mathcal{P}^{-1/2} y_*$. Using a backward transformation from the y_k in the CG method (for details we refer, e.g., to [74]), one can obtain a sequence of approximate solutions x_k to the solution of the original system (6.24). Formally, this approach corresponds to solving the system $\mathcal{P}^{-1} \mathcal{A} x = \mathcal{P}^{-1} b$ or $\mathcal{A} \mathcal{P}^{-1} z = b$, where $x = \mathcal{P}^{-1} z$ (again, the details can be found in [74]). Note that $\mathcal{P}^{-1} \mathcal{A}$ and $\mathcal{A} \mathcal{P}^{-1}$ are generally nonsymmetric, and the equivalence is possible only due to the transformation (6.43). Ideally, we seek a symmetric positive preconditioner $\mathcal{P}$ that is spectrally equivalent to the system matrix $\mathcal{A}$, i.e., there exist positive constants $0 < \hat{\beta} \leq \hat{\alpha}$ such that for every nonzero vector $x \neq 0$, we have $\hat{\beta} \leq \frac{(\mathcal{A}x,x)}{(\mathcal{P}x,x)} \leq \hat{\alpha}$ over the class of given problems with increasing dimension. For the relative error in the preconditioned CG method (measured again by the $\mathcal{A}$-norm), it follows from (6.30) that

$$\frac{\|x - x_k\|_{\mathcal{A}}}{\|x - x_0\|_{\mathcal{A}}} \leq 2 \left(\frac{\sqrt{\kappa(\mathcal{P}^{-1/2}\mathcal{A}\mathcal{P}^{-1/2})} - 1}{\sqrt{\kappa(\mathcal{P}^{-1/2}\mathcal{A}\mathcal{P}^{-1/2})} + 1} \right)^k \leq 2 \left(\frac{\sqrt{\hat{\alpha}} - \sqrt{\hat{\beta}}}{\sqrt{\hat{\alpha}} + \sqrt{\hat{\beta}}} \right)^k . \qquad (6.44)$$

If the constants $\hat{\alpha}$ and $\hat{\beta}$ are independent of the matrix dimension, then also the bound for the relative error in CG does not depend on the matrix dimension. Alternatively, we say that $\mathcal{P}$ is norm equivalent to $\mathcal{A}$, if there exist positive constants $0 < \hat{\beta} \leq \hat{\alpha}$ such that for every nonzero vector $x \neq 0$, we have $\hat{\beta} \leq \frac{\|\mathcal{A}x\|}{\|\mathcal{P}x\|} \leq \hat{\alpha}$ over the class of given problems with increasing dimension. Note that the condition number of $\mathcal{A}\mathcal{P}^{-1}$ satisfies then the bound

$$\kappa(\mathcal{A}\mathcal{P}^{-1}) = \|\mathcal{A}\mathcal{P}^{-1}\| \|\mathcal{P}\mathcal{A}^{-1}\| \leq \hat{\alpha}/\hat{\beta},$$

but it is different from the condition number $\kappa(\mathcal{P}^{-1/2}\mathcal{A}\mathcal{P}^{-1/2})$. Indeed, due to the similarity of matrices $\mathcal{A}\mathcal{P}^{-1}$ and $\mathcal{P}^{-1/2}\mathcal{A}\mathcal{P}^{-1/2}$, their eigenvalues coincide, but $\mathcal{A}\mathcal{P}^{-1}$ is in general nonsymmetric. Therefore its extremal singular values and eigenvalues can differ. The concepts of norm and spectral equivalence are in general not equivalent. For a generalization of these results to infinite-dimensional operators, we refer to [28].

Symmetric indefinite system + symmetric positive definite preconditioner The indefinitness of the saddle point problem (1.1) brings into the field of preconditioning a completely new element. If the matrix $\mathcal{A} = \mathbb{A}$ is symmetric but indefinite, it is not entirely clear what properties should the preconditioning matrix $\mathcal{P}$ have. If $\mathcal{P}$ is symmetric positive definite, then the transformation $\mathcal{P}^{-1/2}\mathcal{A}\mathcal{P}^{-1/2}$ leads again to preconditioned system (6.43) that is symmetric indefinite. In this case the method of choice for solving (6.43) is most often the MINRES method [64]. Note that similar backward transformations as in the positive definite case can be used (although they are less straightforward) in MINRES. For details we refer to [82]. With a symmetric and positive definite preconditioner, we have the following bound for the relative residual norm in the preconditioned MINRES

$$\frac{\|r_k\|_{\mathcal{P}^{-1}}}{\|r_0\|_{\mathcal{P}^{-1}}} \leq \min_{\substack{p \in P_k \\ p(0)=1}} \max_{\lambda \in \mathrm{sp}(\mathcal{P}^{-1}\mathcal{A})} |p(\lambda)| \leq \min_{\substack{p \in P_k \\ p(0)=1}} \max_{\lambda \in [-\hat{\alpha}, -\hat{\beta}] \cup [\hat{\gamma}, \hat{\delta}]} |p(\lambda)|, \qquad (6.45)$$

where $[-\hat{\alpha}, -\hat{\beta}] \cup [\hat{\gamma}, \hat{\delta}]$ is an inclusion set for all of the eigenvalues of $\mathcal{P}^{-1}\mathcal{A}$; see [81] or [82].

Symmetric indefinite system + symmetric indefinite preconditioner If the preconditioner $\mathcal{P}$ is symmetric indefinite, then its square root does not exist in real arithmetics, and the preconditioned system (6.43) that preserves the symmetry does not have a meaning. Thus in the saddle-point context, the preconditioning of the symmetric indefinite matrix $\mathcal{A} = \mathbb{A}$ by the symmetric indefinite preconditioner

$\mathcal{P}$ gives preconditioned matrices $\mathcal{P}^{-1}\mathcal{A}$ or $\mathcal{A}\mathcal{P}^{-1}$. Surprisingly, this leads to solution of nonsymmetric systems (6.40) and (6.41), respectively. A natural question that arises here is which nonsymmetric Krylov subspace method should be applied. Since both $\mathcal{A}$ and $\mathcal{P}$ are symmetric, it appears that the matrix $\mathcal{A}\mathcal{P}^{-1}$ is $\mathcal{H}$-symmetric, i.e., it satisfies the relation

$$(\mathcal{A}\mathcal{P}^{-1})^T \mathcal{H} = \mathcal{H}(\mathcal{A}\mathcal{P}^{-1}), \tag{6.46}$$

where the matrix $\mathcal{H}$ is defined as $\mathcal{H} = \mathcal{P}^{-1}$. Similarly, the matrix $\mathcal{P}^{-1}\mathcal{A}$ satisfies the relation

$$(\mathcal{P}^{-1}\mathcal{A})^T \mathcal{H} = \mathcal{H}(\mathcal{P}^{-1}\mathcal{A}) \tag{6.47}$$

and so $\mathcal{P}^{-1}\mathcal{A}$ is $\mathcal{H}$-symmetric with $\mathcal{H}$ given as $\mathcal{H} = \mathcal{P}$. In both cases, as a natural method of choice seems to be the $\mathcal{H}$-symmetric variant of Bi-CG or QMR [31]. Note that due to the "peak/plateau" relation (6.39) between $\mathcal{H}$-symmetric Bi-CG and $\mathcal{H}$-symmetric QMR methods, there will be no significant difference in practical behavior of these two methods. Both these two methods are based on simplified $\mathcal{H}$-symmetric Lanczos process for computing the biorthogonal bases $\mathcal{V}_k$ and $\mathcal{W}_k$ spanning the subspaces $\mathcal{K}_k(\mathcal{A}\mathcal{P}^{-1}, r_0)$ and $\mathcal{K}_k(\mathcal{P}^{-1}\mathcal{A}, r_0)$, respectively. In the nonsymmetric Lanczos algorithm, these basis vectors are computed by means of two three-term recurrences. The main point here is that due to (6.46) we can show that $\mathcal{W}_k = \mathcal{H}\mathcal{V}_k$, and thus due to $\mathcal{K}_k(\mathcal{P}^{-1}\mathcal{A}, r_0) = \mathcal{H}\mathcal{K}_k(\mathcal{A}\mathcal{P}^{-1}, r_0)$, one must to compute only the basis $\mathcal{V}_k$, and the biorthogonalization algorithm is significantly simplified. It follows from [71] that $\mathcal{H}$-symmetric Bi-CG algorithm is formally nothing but classical CG algorithm preconditioned with indefinite matrix $\mathcal{H}$. Indeed, the preconditioned conjugate gradient method applied to indefinite system (6.24) where $\mathcal{A} = \mathbb{A}$ with the symmetric indefinite preconditioner $\mathcal{P}$ actually corresponds to the conjugate gradient method applied to nonsymmetric (and often nonnormal) preconditioned system (6.41) with $\mathbb{A}\mathbb{P}^{-1}$. Nevertheless, due to particular structure of $\mathbb{A}\mathbb{P}^{-1}$ and due to particular structure of right-hand side vectors, the convergence of the CG method can be theoretically justified in some cases (see, e.g., the case with constraint preconditioning discussed in Chap. 8 or in [71]).

The concept of looking for bilinear forms or nonstandard inner products defined by the matrix $\mathcal{H}$ such that the matrix $\mathcal{A}\mathcal{P}^{-1}$ satisfies (6.46) was further developed in [79] where general conditions for self-adjointness of $\mathcal{A}\mathcal{P}^{-1}$ were considered and a number of existing and new examples were given.

Symmetric indefinite system + nonsymmetric preconditioner Although $\mathcal{A}$ is symmetric indefinite, it is clear that if the preconditioning matrix $\mathcal{P}$ is nonsymmetric, then the preconditioned systems (6.40) and (6.41) are nonsymmetric, and one must apply nonsymmetric iterative method. Despite its computational cost, the most frequently used and theoretically analyzed is the GMRES method [75]. Its practical use is justified due to the fact that the method with efficient preconditioning often converges very quickly and its approximate solution reaches a desired tolerance level much earlier than it becomes too expensive. Some GMRES convergence analyses in applications are based on the field of values of the preconditioned matrix $\mathcal{A}\mathcal{P}^{-1}$ and on using the bound (6.38). Ideally one looks for a preconditioning matrix $\mathcal{P}$, which is equivalent to the matrix $\mathcal{A}$ with respect to the field of values, i.e., there exist positive constants $0 < \hat{\beta} \leq \hat{\alpha}$ such that for every nonzero vector $x \neq 0$, we have $\hat{\beta}\|x\|^2 \leq (\mathcal{A}\mathcal{P}^{-1}x, x)$ and $(\mathcal{A}\mathcal{P}^{-1}x, x) \leq \hat{\alpha}\|x\|^2$ over a given class of problems with increasing dimension. The relative residual norm in preconditioned GMRES can be then bounded as

$$\frac{\|r_k\|}{\|r_0\|} \leq \left[1 - \left(\frac{\min_x(\mathcal{A}\mathcal{P}^{-1}x, x)}{\max_x(\mathcal{A}\mathcal{P}^{-1}x, x)}\right)^2\right]^{\frac{k}{2}} \leq \left[1 - \left(\frac{\hat{\beta}}{\hat{\alpha}}\right)^2\right]^{\frac{k}{2}}. \tag{6.48}$$

If $\hat{\alpha}$ and $\hat{\beta}$ are constants independent of the matrix dimension, then the bound for the convergence rate of preconditioned GMRES will not be dependent on this dimension. In addition, if $\hat{\alpha}$ is close to $\hat{\beta}$, then a fast convergence of GMRES can be expected. This approach is mainly used for inexact versions of block-diagonal or block-triangular preconditioners, where one of their diagonal blocks is negative definite. Another possible approach is based on the analysis of a degree of the minimal polynomial for the matrix $\mathcal{A}\mathcal{P}^{-1}$. It appears that this degree can be sometimes very small (it can be even equal to 2 or 3). Then one can expect that every Krylov subspace method will terminate within this (small) number of steps. This property holds for the so-called "exact" versions of block-diagonal or block-triangular preconditioners, or with some modification also for the constraint preconditioner, which will be discussed in next chapter.

6.4   Multigrid Methods

Multigrid methods are considered as one of the most efficient solvers for saddle-point problems arising from discretizations of partial differential equations. Their construction is therefore naturally tied to the properties of continuous problem that actually leads to a sequence of saddle-point problems $\mathbb{A}_h \mathbb{x}_h = \mathbb{b}_h$, where h denotes a parameter that gives a measure for the size of the grid used in the discretization. It is clear that if we go from coarser grids to finer grids, the dimension of these saddle-point problems increases with the decreasing h. As the saddle-point problem corresponding to the problem on coarser grids is usually simpler to solve than finer grids, the multigrid methods combine stationary iterative methods to smooth the error of approximate solutions on finer grids with the corrections computed on coarser grids. Since the main focus of this textbook is on linear algebra, in this section, we will restrict ourselves only to the description of the V-cycle multigrid algorithm applied to the preconditioned system

$$\mathbb{A}_h \mathbb{P}_h^{-1} \mathbb{y}_h = \mathbb{b}_h, \quad \mathbb{x}_h = \mathbb{P}_h^{-1} \mathbb{y}_h, \tag{6.49}$$

where $\mathbb{P}_h$ denotes the preconditioner for the problem on the grid of the size h. An essential component of multigrid methods is an inexpensive smoothing procedure, which consists of a few steps of stationary iterative method applied to (6.49) in the form

$$x_{h,k+1} = x_{h,k} + \mathbb{P}_h^{-1}\mathbb{M}_h^{-1}(\mathbb{b}_h - \mathbb{A}_h x_{h,k}), \tag{6.50}$$

with the appropriate splitting matrix $\mathbb{M}_h$ satisfying $\mathbb{A}_h\mathbb{P}_h^{-1} = \mathbb{M}_h - \mathbb{N}_h$ for some $\mathbb{N}_h$ and with initial guess $x_{h,0}$. If the oscillating components of the error are damped, the second step is to look for the correction on the coarser grid of the size $2h$ and approximate the remaining smooth part, solving the correction problem $\mathbb{A}_{2h}x_{2h} = r_{2h}$, where $r_{2h} = \mathbb{R}_{h,2h}(\mathbb{b}_h - \mathbb{A}_h x_{h,k+1})$ is the restricted residual and where the term $\mathbb{R}_{h,2h}$ denotes the restriction operator from a fine grid to a course grid. The solution of the course problem is prolonged using the prolongation operator $\mathbb{P}_{2h,h}$ (usually $\mathbb{P}_{2h,h} = \mathbb{R}_{h,2h}^T$) and added to the approximate solution. Finally, the correction step is followed by several post-smoothing steps of stationary iterative method (6.50). We continue the whole procedure recursively and double the grid size until we get a sufficiently small saddle-point problem that can be solved using a direct method. One step of V-cycle multigrid method is summarized in Algorithm 6.3.

Algorithm 6.3 V-cycle multigrid method

---

solve $\mathbb{A}_h x_h = \mathbb{b}_h$ with the initial guess $x_{h,0}$

for $k = 0, 1, \ldots, \ell - 1$

    $x_{h,k+1} = x_{h,k} + \mathbb{P}_h^{-1}\mathbb{M}_h^{-1}(\mathbb{b}_h - \mathbb{A}_h x_{h,k})$

end

$r_{2h} = \mathbb{R}_{h,2h}(\mathbb{b}_h - \mathbb{A}_h x_{h,\ell})$

if the grid is not course enough, then

    solve recursively $\mathbb{A}_{2h}x_{2h} = r_{2h}$ with the initial guess $x_{2h,0} = 0$

else

    solve $\mathbb{A}_{2h}x_{2h} = r_{2h}$ directly

end if

$x_{h,0} = x_{h,\ell} + \mathbb{P}_{2h,h}x_{2h}$

for $k = 0, 1, \ldots, \ell - 1$

    $x_{h,k+1} = x_{h,k} + \mathbb{P}_h^{-1}\mathbb{M}_h^{-1}(\mathbb{b}_h - \mathbb{A}_h x_{h,k})$

end

A key point of an efficient multigrid method is the choice of the smoothing procedure with the task to smoothen the error on each of a hierarchy of grids which is essential for the convergence. Several classes of smoothers have been proposed and analyzed for saddle-point problems that arise from discretizations of differential equations, including the inexact Uzawa's methods that are based on the block-triangular splitting (6.5) and stationary methods that are based on symmetric indefinite splitting (6.12). These two main classes of methods were thoroughly analyzed in [85] in the context of preconditioned saddle-point problems; see also the discussion in Sect. 6.1. The analysis of smoothing properties of inexact Uzawa's methods and stationary methods with symmetric indefinite splittings can be found in [15, 19, 84]. The efficiency of multigrid methods for solving saddle-point problems arising from the discretization of the Stokes equations, a fundamental application in computational fluid dynamics, was tested and confirmed in [21]; see also [22].

Chapter 7
Preconditioners for Saddle-Point Problems

Preconditioners for saddle-point problems exploit the 2-by-2 block structure of the saddle-point matrix $\mathbb{A}$. They have been a subject of active research in a whole range of applications. For a current survey, we refer to [67]. Preconditioning techniques for saddle-point systems include the following classes of schemes:

- Each splitting $\mathbb{A} = \mathbb{M} - \mathbb{N}$ from the stationary iterative method can be used for preconditioning of a saddle-point system. Since the splitting matrix $\mathbb{M}$ is usually chosen to be easily invertible, or at least the system with the matrix $\mathbb{M}$ is easy to solve, the same applies also to the preconditioner matrix $\mathbb{P} = \mathbb{M}$.
- Each incomplete factorization for symmetric indefinite matrix such as $\mathbb{A} = \mathbb{L}\mathbb{D}\mathbb{L}^T$ or $\mathbb{A} = \mathbb{L}\mathbb{T}\mathbb{L}^T$ can be used to precondition the system with the saddle-point matrix $\mathbb{A}$. Usually it represents an exact factorization of a nearby matrix $\hat{\mathbb{A}} \approx \mathbb{A}$, and we can take $\mathbb{P} = \hat{\mathbb{A}}$. Usually it is assumed that $\hat{\mathbb{A}}$ is given in the factorized form $\hat{\mathbb{A}} = \hat{\mathbb{L}}\hat{\mathbb{D}}\hat{\mathbb{L}}^T$ or $\hat{\mathbb{A}} = \hat{\mathbb{L}}\hat{\mathbb{T}}\hat{\mathbb{L}}^T$, and therefore the inverse of the preconditioner matrix $\mathbb{P}$ is also given in the factorized form and it is easily computable.
- Block diagonal and triangular preconditioners are directly based on the block structure of $\mathbb{A}$ approximating its diagonal or triangular part

$$\mathbb{P} = \begin{pmatrix} A & 0 \\ 0 & \pm B^T A^{-1} B \end{pmatrix}, \quad \mathbb{P} = \begin{pmatrix} A & B \\ 0 & \pm B^T A^{-1} B \end{pmatrix}$$

and their inexact versions that approximate diagonal blocks with symmetric matrices $\hat{A} \approx A$ and $\hat{C} \approx B^T A^{-1} B$ as

$$\hat{\mathbb{P}} = \begin{pmatrix} \hat{A} & 0 \\ 0 & \pm \hat{C} \end{pmatrix}, \quad \hat{\mathbb{P}} = \begin{pmatrix} \hat{A} & B \\ 0 & \pm \hat{C} \end{pmatrix}.$$

- Constraint preconditioners which represent a (potentially cheaper) solution of a saddle-point system with the same constraints as in the original saddle-point

© Springer Nature Switzerland AG 2018
M. Rozložník, *Saddle-Point Problems and Their Iterative Solution*,
Nečas Center Series, https://doi.org/10.1007/978-3-030-01431-5_7

system

$$\mathbb{P} = \begin{pmatrix} D & B \\ B^T & -E \end{pmatrix} = \begin{pmatrix} I & 0 \\ B^T D^{-1} & I \end{pmatrix} \begin{pmatrix} D & 0 \\ 0 & -E - B^T D^{-1} B \end{pmatrix} \begin{pmatrix} I & D^{-1} B \\ 0 & I \end{pmatrix}.$$

7.1 Block Diagonal and Triangular Preconditioners

Block diagonal preconditioners This class of preconditioners heavily relies on the availability of a cheap solution of systems with matrices A and $B^T A^{-1} B$. If A is symmetric positive definite and B has a full-column rank, then $B^T A^{-1} B$ is symmetric positive definite, so as the exact block diagonal preconditioner

$$\mathbb{P} = \begin{pmatrix} A & 0 \\ 0 & B^T A^{-1} B \end{pmatrix}. \tag{7.1}$$

Considering (1.3) and (7.1), the preconditioned matrices $\mathbb{P}^{-1}\mathbb{A}$ and $\mathbb{A}\mathbb{P}^{-1}$ are nonsingular but nonsymmetric, and they are given as

$$\mathbb{A}\mathbb{P}^{-1} = \begin{pmatrix} I & B(B^T A^{-1} B)^{-1} \\ B^T A^{-1} & 0 \end{pmatrix} = (\mathbb{P}^{-1}\mathbb{A})^T. \tag{7.2}$$

It follows from the papers [61] and [44] that the matrix $\mathbb{A}\mathbb{P}^{-1}$ is diagonalizable and it satisfies the identity

$$(\mathbb{A}\mathbb{P}^{-1})^2 - \mathbb{A}\mathbb{P}^{-1} = \begin{pmatrix} B(B^T A^{-1} B)^{-1} B^T A^{-1} & 0 \\ 0 & I \end{pmatrix}. \tag{7.3}$$

Because the matrix $B(B^T A^{-1} B)^{-1} B^T A^{-1}$ represents a projector, it is clear from (7.3) that

$$\left((\mathbb{A}\mathbb{P}^{-1})^2 - \mathbb{A}\mathbb{P}^{-1} \right)^2 = (\mathbb{A}\mathbb{P}^{-1})^2 - \mathbb{A}\mathbb{P}^{-1}. \tag{7.4}$$

Since $\mathbb{A}\mathbb{P}^{-1}$ is nonsingular, from (7.4) we have that

$$\left(\mathbb{A}\mathbb{P}^{-1} - \mathbb{I} \right) \left[(\mathbb{A}\mathbb{P}^{-1})^2 - \mathbb{A}\mathbb{P}^{-1} - \mathbb{I} \right] = 0. \tag{7.5}$$

Consequently, $\mathbb{A}\mathbb{P}^{-1}$ and $\mathbb{P}^{-1}\mathbb{A}$ have only three different nonzero eigenvalues

$$\mathrm{sp}(\mathbb{A}\mathbb{P}^{-1}) = \left\{ 1, \tfrac{1}{2}(1 \pm \sqrt{5}) \right\}.$$

Therefore, the minimum polynomial degree is 3. This means that any Krylov subspace method, including the GMRES method, will terminate in at most three steps for any initial guess x_0 [61].

If we assume again that A is symmetric positive definite and B has a full-column rank, then the block diagonal preconditioner

$$\mathbb{P} = \begin{pmatrix} A & 0 \\ 0 & -B^T A^{-1} B \end{pmatrix} \tag{7.6}$$

is nonsingular but symmetric indefinite. The preconditioned matrices $\mathbb{A}\mathbb{P}^{-1}$ and $\mathbb{P}^{-1}\mathbb{A}$ satisfy

$$\mathbb{A}\mathbb{P}^{-1} = \begin{pmatrix} I & -B(B^T A^{-1} B)^{-1} \\ B^T A^{-1} & 0 \end{pmatrix} = (\mathbb{P}^{-1}\mathbb{A})^T. \tag{7.7}$$

Similarly to the previous case, the preconditioned matrix $\mathbb{A}\mathbb{P}^{-1}$ has a minimal polynomial of degree 3 due to the identity

$$\left(\mathbb{P}^{-1}\mathbb{A} - \mathbb{I}\right)\left[(\mathbb{P}^{-1}\mathbb{A})^2 - (\mathbb{P}^{-1}\mathbb{A}) + \mathbb{I}\right] = 0 \tag{7.8}$$

and we get a termination in at most three steps of any Krylov subspace method.

The application of the exact preconditioners (7.1) and (7.6) is expensive, and in practice they are applied inexactly. Instead of (7.1) one can use the inexact preconditioner

$$\hat{\mathbb{P}} = \begin{pmatrix} \hat{A} & 0 \\ 0 & \hat{C} \end{pmatrix}, \tag{7.9}$$

where $\hat{A}$ is a symmetric positive definite approximation to the symmetric positive definite matrix block A and $\hat{C}$ is a symmetric positive definite approximation to the symmetric positive definite matrix $B^T A^{-1} B$. Since in practice the exact inverse of A is not available, very often $\hat{C}$ is a symmetric positive approximation to the matrix $B^T \hat{A}^{-1} B$ that is also symmetric positive definite. The matrix $\hat{\mathbb{P}}$ is thus also symmetric positive definite, and we can consider the two-sided preconditioning of the saddle-point system

$$\hat{\mathbb{P}}^{-1/2}\mathbb{A}\hat{\mathbb{P}}^{-1/2} = \begin{pmatrix} \hat{A}^{-1/2} & 0 \\ 0 & \hat{C}^{-1/2} \end{pmatrix} \begin{pmatrix} A & B \\ B^T & 0 \end{pmatrix} \begin{pmatrix} \hat{A}^{-1/2} & 0 \\ 0 & \hat{C}^{-1/2} \end{pmatrix}$$

$$= \begin{pmatrix} \hat{A}^{-1/2} A \hat{A}^{-1/2} & \hat{A}^{-1/2} B \hat{C}^{-1/2} \\ \hat{C}^{-1/2} B^T \hat{A}^{-1/2} & 0 \end{pmatrix} = \begin{pmatrix} \tilde{A} & \tilde{B} \\ \tilde{B}^T & 0 \end{pmatrix}. \tag{7.10}$$

In practical computations, we look for a symmetric positive definite approximation $\hat{A}$ that is spectrally equivalent to the block A, i.e., there exist positive constants

$0 < \hat{\beta} \le \hat{\alpha}$ such that

$$\hat{\beta}(\hat{A}x, x) \le (Ax, x) \le \hat{\alpha}(\hat{A}x, x)$$

for all vectors $x \in \mathcal{R}^m$. Similarly, we look for a symmetric positive definite approximation $\hat{C}$ that is spectrally equivalent to the matrix $B^T \hat{A}^{-1} B$, i.e., there exist positive constants $0 < \hat{\gamma} \le \hat{\delta}$ such that

$$\hat{\gamma}(\hat{C}y, y) \le (B^T \hat{A}^{-1} By, y) \le \hat{\delta}(\hat{C}y, y)$$

for all vectors $y \in \mathcal{R}^n$. Ideally, if all constants do not depend on the dimension of the system, then the spectrum of the preconditioned matrix $\hat{\mathbb{P}}^{-1/2} \mathbb{A} \hat{\mathbb{P}}^{-1/2}$ is also independent of the dimension, and it determines the rate of convergence of iterative methods such as the MINRES method that is then uniformly bounded over the class of systems with increasing dimension arising in applications such as partial differential equations discretized with a uniformly refined mesh.

We can distinguish between two extremal cases. In the first case we assume that the block A is approximated exactly with $\hat{A} = A$, while the negative of the Schur complement matrix is approximated with $\hat{C} \approx B^T A^{-1} B$. Then

$$\hat{\mathbb{P}}^{-1/2} \mathbb{A} \hat{\mathbb{P}}^{-1/2} = \begin{pmatrix} I & A^{-1/2} B \hat{C}^{-1/2} \\ \hat{C}^{-1/2} B^T A^{-1/2} & 0 \end{pmatrix} = \begin{pmatrix} I & \hat{B} \\ \hat{B}^T & 0 \end{pmatrix}.$$

Since from Proposition 3.4 the spectrum of $\hat{\mathbb{P}}^{-1/2} \mathbb{A} \hat{\mathbb{P}}^{-1/2}$ is determined by the singular values of $\hat{B}$, so essential here is the spectrum of the matrix $\hat{B}^T \hat{B} = \hat{C}^{-1/2} B^T A^{-1} B \hat{C}^{-1/2}$. Indeed, we have

$$\hat{\gamma} \|y\|^2 \le (\hat{B}^T \hat{B}y, y) = (\hat{C}^{-1/2} B^T \hat{A}^{-1} B \hat{C}^{-1/2} y, y) \le \hat{\delta} \|y\|^2$$

for all vectors $y \in \mathcal{R}^n$. In the second case, we assume the approximation of the block A with $\hat{A} \approx A$, while $\hat{C}$ exactly satisfies the identity $\hat{C} = B^T \hat{A}^{-1} B$. Consequently,

$$\hat{\mathbb{P}}^{-1/2} \mathbb{A} \hat{\mathbb{P}}^{-1/2} = \begin{pmatrix} \hat{A}^{-1/2} A \hat{A}^{-1/2} & \hat{A}^{-1/2} B \hat{C}^{-1/2} \\ \hat{C}^{-1/2} B^T \hat{A}^{-1/2} & 0 \end{pmatrix} = \begin{pmatrix} \hat{A}^{-1/2} A \hat{A}^{-1/2} & \hat{B} \\ \hat{B}^T & 0 \end{pmatrix},$$

whereas $\hat{B}^T \hat{B} = \hat{C}^{-1/2} B^T \hat{A}^{-1/2} \hat{A}^{-1/2} B \hat{C}^{-1/2} = \hat{C}^{-1/2} B^T \hat{A}^{-1} B \hat{C}^{-1/2} = I$. The matrix $\hat{B}$ has orthonormal columns with $\kappa(\hat{B}) = 1$, and from Proposition 3.5 the spectrum of $\hat{\mathbb{P}}^{-1/2} \mathbb{A} \hat{\mathbb{P}}^{-1/2}$ is determined by the spectrum of the matrix $\hat{A}^{-1/2} A \hat{A}^{-1/2}$. Indeed, we have that

$$\hat{\beta} \|x\|^2 \le (\hat{A}^{-1/2} A \hat{A}^{-1/2} x, x) \le \hat{\alpha} \|x\|^2$$

for all vectors $x \in \mathcal{R}^m$. Optimal block diagonal preconditioners for several classes of saddle-point problems that use the multigrid techniques have been proposed in [76] or [49].

Block triangular preconditioners These preconditioners are naturally nonsymmetric and again rely on the availability of a cheap approximate solution of systems with A and $B^T A^{-1} B$. The exact block triangular preconditioner is given as

$$\mathbb{P} = \begin{pmatrix} A & B \\ 0 & B^T A^{-1} B \end{pmatrix}. \tag{7.11}$$

The preconditioned matrix $\mathbb{AP}^{-1}$ has then the form

$$\mathbb{AP}^{-1} = \begin{pmatrix} A & B \\ B^T & 0 \end{pmatrix} \begin{pmatrix} A^{-1} & -A^{-1}B(B^T A^{-1} B)^{-1} \\ 0 & (B^T A^{-1} B)^{-1} \end{pmatrix} = \begin{pmatrix} I & 0 \\ B^T A^{-1} & -I \end{pmatrix}. \tag{7.12}$$

The matrix $\mathbb{AP}^{-1}$ is diagonalizable and its spectrum is $\mathrm{sp}(\mathbb{AP}^{-1}) = \{\pm 1\}$. In addition, $\mathbb{AP}^{-1}$ has the minimal polynomial

$$(\mathbb{AP}^{-1} - I)(\mathbb{AP}^{-1} + I) = 0. \tag{7.13}$$

Changing the sign of the second diagonal block in (7.11), we get the preconditioner

$$\mathbb{P} = \begin{pmatrix} A & B \\ 0 & -B^T A^{-1} B \end{pmatrix} \tag{7.14}$$

and the preconditioned matrix $\mathbb{AP}^{-1}$ in the form

$$\mathbb{AP}^{-1} = \begin{pmatrix} A & B \\ B^T & 0 \end{pmatrix} \begin{pmatrix} A^{-1} & -A^{-1}B(B^T A^{-1} B)^{-1} \\ 0 & (B^T A^{-1} B)^{-1} \end{pmatrix} = \begin{pmatrix} I & 0 \\ B^T A^{-1} & I \end{pmatrix}. \tag{7.15}$$

The preconditioned matrix $\mathbb{AP}^{-1}$ has then the spectrum $\mathrm{sp}(\mathbb{AP}^{-1}) = \{1\}$, but it is not diagonalizable. The minimal polynomial degree is 2 due to

$$(\mathbb{AP}^{-1} - I)^2 = 0. \tag{7.16}$$

Consequently any converging Krylov subspace method such as GMRES terminates in at most two steps [44, 61]. As it was shown in [44], this is true not only for $\mathbb{A}$ in the saddle-point form (1.3) but for the case of $\mathbb{A}$ with the 2-by-2 block structure (1.2), where $C \neq 0$. In this case, however, the diagonal block $B^T A^{-1} B$ in (7.11) must be replaced by $B^T A^{-1} B - C$. Similarly $-B^T A^{-1} B$ in (7.14) must be replaced by the Schur complement matrix $C - B^T A^{-1} B$.

The exact application of both block triangular preconditioners is expensive, and in practical situations their inexact versions

$$\hat{\mathbb{P}} = \begin{pmatrix} \hat{A} & B \\ 0 & \pm\hat{C} \end{pmatrix} \tag{7.17}$$

are used. The application of $\hat{\mathbb{P}}^{-1}$ is usually performed in the factorized form

$$\hat{\mathbb{P}}^{-1} = \begin{pmatrix} \hat{A}^{-1} & 0 \\ 0 & I \end{pmatrix} \begin{pmatrix} I & -B \\ 0 & I \end{pmatrix} \begin{pmatrix} I & 0 \\ 0 & \pm\hat{C}^{-1} \end{pmatrix} \tag{7.18}$$

and it is based on a cheap solution of systems with symmetric positive definite matrices $\hat{A} \approx A$ and $\hat{C} \approx B^T A^{-1} B$. The resulting preconditioned systems are nonsymmetric, as

$$\mathbb{A}\hat{\mathbb{P}}^{-1} = \begin{pmatrix} A & B \\ B^T & 0 \end{pmatrix} \begin{pmatrix} \hat{A}^{-1} & \mp\hat{A}^{-1}B\hat{C}^{-1} \\ 0 & \hat{C}^{-1} \end{pmatrix} = \begin{pmatrix} A\hat{A}^{-1} & (I \mp A\hat{A}^{-1})B\hat{C}^{-1} \\ B^T\hat{A}^{-1} & \mp B^T\hat{A}^{-1}B\hat{C}^{-1} \end{pmatrix}.$$

The matrices $\hat{A}$ and $\hat{C}$ are often required to satisfy

$$\hat{\beta}(\hat{A}x, x) \leq (Ax, x) \leq \hat{\alpha}(\hat{A}x, x) \tag{7.19}$$

for all vectors $x \in \mathcal{R}^m$, and also

$$\hat{\gamma}(\hat{C}y, y) \leq (B^T\hat{A}^{-1}By, y) \leq \hat{\delta}(\hat{C}y, y) \tag{7.20}$$

for all vectors $y \in \mathcal{R}^n$, where $0 < \hat{\beta} \leq \hat{\alpha}$ and $0 < \hat{\gamma} \leq \hat{\delta}$. Ideally these constants are independent of the system dimension. In the nonsymmetric case, the independence of the spectrum of $\mathbb{A}\hat{\mathbb{P}}^{-1}$ does not guarantee that the convergence rate of Krylov subspace methods will not deteriorate in the class of problems with increasing dimension that come from a uniformly refined discretization of some partial differential equation. Multigrid and domain decomposition methods are examples of preconditioners that meet the requirements (7.19) or (7.20). Optimal preconditioners of this type for a class of saddle-point problems arising from finite element discretizations were developed and analyzed in [24] or [48].

7.2   Constraint Preconditioners

Constraint preconditioners are constructed so that the preconditioning matrix $\mathbb{P}$ keeps the same 2-by-2 block structure with the same off-diagonal blocks as the original matrix $\mathbb{A}$. As it will be clear below, the Schur complement method for the solution of saddle-point problem can be also seen as an application of the constraint preconditioner in the form

$$\mathbb{P} = \begin{pmatrix} A & B \\ B^T & B^T A^{-1} B - I \end{pmatrix}. \tag{7.21}$$

The block triangular factorization of the matrix $\mathbb{P}$ leads to

$$\mathbb{P} = \begin{pmatrix} I & 0 \\ B^T A^{-1} & I \end{pmatrix} \begin{pmatrix} A & 0 \\ 0 & -I \end{pmatrix} \begin{pmatrix} I & A^{-1} B \\ 0 & I \end{pmatrix} \tag{7.22}$$

and the inverse of $\mathbb{P}$ can be then also easily factorized as

$$\mathbb{P}^{-1} = \begin{pmatrix} I & -A^{-1} B \\ 0 & I \end{pmatrix} \begin{pmatrix} A^{-1} & 0 \\ 0 & -I \end{pmatrix} \begin{pmatrix} I & 0 \\ -B^T A^{-1} & I \end{pmatrix}. \tag{7.23}$$

Provided that A is symmetric positive definite, it follows from (7.22) that $\mathbb{P}$ is symmetric indefinite with m positive eigenvalues and with n negative eigenvalues. Therefore, the inertia of $\mathbb{P}$ is exactly the same as the inertia of $\mathbb{A}$. The preconditioned matrix $\mathbb{A}\mathbb{P}^{-1}$ is block lower triangular

$$\mathbb{A}\mathbb{P}^{-1} = \begin{pmatrix} I & 0 \\ (I - B^T A^{-1} B) B^T A^{-1} & B^T A^{-1} B \end{pmatrix}, \tag{7.24}$$

and so it is nonsymmetric but diagonalizable with the spectrum satisfying

$$\mathrm{sp}(\mathbb{A}\mathbb{P}^{-1}) \subset \{1\} \cup \mathrm{sp}(B^T A^{-1} B). \tag{7.25}$$

For details we refer, e.g., to [20]. In the following we consider any Krylov subspace method applied to the system (2.20) that takes the initial guess $\mathbb{x}_0 = \begin{pmatrix} x_0 \\ y_0 \end{pmatrix}$ with the residual $\mathbb{r}_0 = \mathbb{b} - \mathbb{A}\mathbb{x}_0$ and generates the approximate solutions $\mathbb{x}_{k+1} = \begin{pmatrix} x_{k+1} \\ y_{k+1} \end{pmatrix}$, $k = 0, 1, \ldots$ with the error vectors $\mathbb{e}_{k+1}$ and residual vectors $\mathbb{r}_{k+1}$ given as

$$\mathbb{e}_{k+1} = \begin{pmatrix} x_* - x_{k+1} \\ y_* - y_{k+1} \end{pmatrix}, \quad \mathbb{r}_{k+1} = \begin{pmatrix} c \\ d \end{pmatrix} - \begin{pmatrix} A & B \\ B^T & 0 \end{pmatrix} \begin{pmatrix} x_{k+1} \\ y_{k+1} \end{pmatrix}. \tag{7.26}$$

We assume the right preconditioning of $\mathbb{A}\mathbb{x} = \mathbb{b}$ with the preconditioner $\mathbb{P}$ leading to the preconditioned system

$$\mathbb{A}\mathbb{P}^{-1}\mathbb{z} = \mathbb{b}, \quad \mathbb{x} = \mathbb{P}^{-1}\mathbb{z}. \tag{7.27}$$

The approximate solutions $\mathbb{x}_{k+1}$ and the residuals $\mathbb{r}_{k+1} = \mathbb{b} - \mathbb{A}\mathbb{x}_{k+1}$ then satisfy

$$\mathbb{x}_{k+1} \in \mathbb{x}_0 + \mathbb{P}^{-1}\mathcal{K}_{k+1}(\mathbb{A}\mathbb{P}^{-1}, \mathbb{r}_0), \quad \mathbb{r}_{k+1} \in \mathbb{r}_0 + \mathbb{A}\mathbb{P}^{-1}\mathcal{K}_{k+1}(\mathbb{A}\mathbb{P}^{-1}, \mathbb{r}_0),$$
$$(7.28)$$

where $\mathcal{K}_{k+1}(\mathbb{A}\mathbb{P}^{-1}, \mathbb{r}_0) = \text{span}(\mathbb{r}_0, \mathbb{A}\mathbb{P}^{-1}\mathbb{r}_0, \ldots, (\mathbb{A}\mathbb{P}^{-1})^k\mathbb{r}_0)$. It is also clear from (7.24) that if one chooses the initial guess $\mathbb{x}_0$ so that $\mathbb{r}_0 = \begin{pmatrix} 0 \\ r_0 \end{pmatrix}$, then due to (7.28) all residual vectors $\mathbb{r}_{k+1}$ will have the same structure

$$\mathbb{r}_{k+1} = \begin{pmatrix} 0 \\ r_{k+1} \end{pmatrix}. \tag{7.29}$$

The identity (7.29) indicates that the components of the approximate solution $\mathbb{x}_{k+1}$ satisfy $Ax_{k+1} + By_{k+1} = c$ and thus they represent approximate solutions from the Schur complement method, characterized by (4.5). Indeed, if the residual vectors satisfy $\mathbb{r}_{k+1} = p_{k+1}(\mathbb{A}\mathbb{P}^{-1})\mathbb{r}_0$ for some polynomial p_{k+1} of degree $k + 1$, then $r_{k+1} = p_{k+1}(B^T A^{-1} B)r_0$, whereas $r_0 = d - B^T x_0 = d - B^T A^{-1}(c - By_0) = d - B^T A^{-1}c + B^T A^{-1}By_0$.

In practical situations, we approximate the diagonal block A with some approximation $\hat{A} \approx A$ and use the inexact preconditioner

$$\hat{\mathbb{P}} = \begin{pmatrix} \hat{A} & B \\ B^T & B^T\hat{A}^{-1}B - I \end{pmatrix} = \begin{pmatrix} I & 0 \\ B^T\hat{A}^{-1} & I \end{pmatrix} \begin{pmatrix} \hat{A} & 0 \\ 0 & -I \end{pmatrix} \begin{pmatrix} I & \hat{A}^{-1}B \\ 0 & I \end{pmatrix}. \tag{7.30}$$

If A is in addition symmetric positive definite, then $\hat{\mathbb{P}}$ is symmetric indefinite with the same inertia as $\mathbb{A}$, and it can be factorized as

$$\hat{\mathbb{P}} = \begin{pmatrix} \hat{A}^{1/2} & 0 \\ B^T\hat{A}^{-1/2} & I \end{pmatrix} \begin{pmatrix} I & 0 \\ 0 & -I \end{pmatrix} \begin{pmatrix} \hat{A}^{1/2} & \hat{A}^{-1/2}B \\ 0 & I \end{pmatrix}.$$

We also consider the solution of saddle-point problem (2.20) via null-space method using the constraint preconditioner

$$\mathbb{P} = \begin{pmatrix} I & B \\ B^T & 0 \end{pmatrix} = \begin{pmatrix} I & 0 \\ B^T & I \end{pmatrix} \begin{pmatrix} I & 0 \\ 0 & -B^T B \end{pmatrix} \begin{pmatrix} I & B \\ 0 & I \end{pmatrix}. \tag{7.31}$$

The inverse of $\mathbb{P}$ can be easily applied using its factorization

$$\mathbb{P}^{-1} = \begin{pmatrix} I & -B \\ 0 & I \end{pmatrix} \begin{pmatrix} I & 0 \\ 0 & -(B^T B)^{-1} \end{pmatrix} \begin{pmatrix} I & 0 \\ -B^T & I \end{pmatrix}. \tag{7.32}$$

In addition, the preconditioned matrix $\mathbb{A}\mathbb{P}^{-1}$ is block upper triangular

$$\mathbb{A}\mathbb{P}^{-1} = \begin{pmatrix} AZZ^T + B(B^T B)^{-1}B^T \ (A - I)B(B^T B)^{-1} \\ 0 \qquad\qquad\qquad I \end{pmatrix}, \tag{7.33}$$

where the matrix $B(B^T B)^{-1}B^T = I - ZZ^T$ represents the orthogonal projector onto $\mathcal{R}(B)$. The matrix $\mathbb{A}\mathbb{P}^{-1}$ is nonsymmetric and non-diagonalizable with the spectrum satisfying

$$\mathrm{sp}(\mathbb{A}\mathbb{P}^{-1}) \subset \{1\} \cup \mathrm{sp}(AZZ^T + B(B^T B)^{-1}B^T)$$
$$\subset \{1\} \cup \mathrm{sp}(ZZ^T AZZ^T) \setminus \{0\}.$$

If the matrix block A is symmetric positive definite, it contains unit eigenvalues and positive eigenvalues of the symmetric positive semi-definite matrix $ZZ^T AZZ^T$, i.e., the eigenvalues of the symmetric positive definite matrix $Z^T AZ$. For details we refer to [55] or [71]. Again, we consider a Krylov subspace method applied to the system (2.20) that takes the initial guess $\mathbb{x}_0 = \begin{pmatrix} x_0 \\ y_0 \end{pmatrix}$ with the residual $\mathbb{r}_0 = \mathbb{b} - \mathbb{A}\mathbb{x}_0$ and generates the approximate solutions $\mathbb{x}_{k+1} = \begin{pmatrix} x_{k+1} \\ y_{k+1} \end{pmatrix}$, $k = 0, 1, \ldots$, with the error vectors $\mathbb{e}_{k+1}$ and residual vectors $\mathbb{r}_{k+1} = \mathbb{b} - \mathbb{A}\mathbb{x}_{k+1}$ given as (7.26). We again assume the right preconditioned system (7.27) and the approximate solutions $\mathbb{x}_{k+1}$ satisfying (7.28). It is clear now from (7.33) that if we choose the initial guess $\mathbb{x}_0$ so that $\mathbb{r}_0 = \begin{pmatrix} r_0 \\ 0 \end{pmatrix}$, then all residual vectors $\mathbb{r}_{k+1}$ will have the same structure

$$\mathbb{r}_{k+1} = \begin{pmatrix} r_{k+1} \\ 0 \end{pmatrix}. \tag{7.34}$$

This shows that the approximate solutions x_{k+1} satisfy the second block equation $B^T x_{k+1} = d$, i.e., they are equal to the approximate solutions (4.16) generated by the null-space method, and they satisfy also

$$B^T (x_* - x_{k+1}) = 0. \tag{7.35}$$

Therefore, the errors $x_* - x_{k+1}$ belong to the null-space $\mathcal{N}(B^T)$. Clearly, if the residual vectors satisfy $\mathbb{r}_{k+1} = p_{k+1}(\mathbb{A}\mathbb{P}^{-1})\mathbb{r}_0$ for some polynomial p_{k+1} of degree $k + 1$, then also

$$r_{k+1} = c - Ax_{k+1} - By_{k+1}$$
$$= p_{k+1}(AZZ^T + B(B^T B)^{-1}B^T)r_0, \tag{7.36}$$

where $r_0 = c - Ax_0 - By_0$ and $B^T x_0 = d$. Consequently, the projection of r_{k+1} onto $\mathcal{N}(B^T)$ is given as

$$ZZ^T r_{k+1} = ZZ^T (c - Ax_{k+1})$$
$$= Zp_{k+1}(Z^T AZ)Z^T (c - Ax_0).$$

In practical situations, provided that B has a full-column rank, we approximate the symmetric positive definite matrix $B^T B$ with the symmetric positive definite matrix $\hat{C} \approx B^T B$ so that we have the block triangular factorization

$$\hat{\mathbb{P}} = \begin{pmatrix} I & 0 \\ B^T & \hat{C}^{1/2} \end{pmatrix} \begin{pmatrix} I & 0 \\ 0 & -I \end{pmatrix} \begin{pmatrix} I & B \\ 0 & \hat{C}^{1/2} \end{pmatrix}. \tag{7.37}$$

The inverse of $\hat{\mathbb{P}}$ is then applied accordingly as

$$\hat{\mathbb{P}}^{-1} = \begin{pmatrix} I & -B\hat{C}^{-1/2} \\ 0 & \hat{C}^{-1/2} \end{pmatrix} \begin{pmatrix} I & 0 \\ 0 & -I \end{pmatrix} \begin{pmatrix} I & 0 \\ \hat{C}^{-1/2}B^T & \hat{C}^{-1/2} \end{pmatrix}.$$

Chapter 8
Numerical Behavior of Saddle-Point Solvers

As we have seen in previous sections, a large amount of work has been devoted to solution techniques for saddle-point problems varying from the fully direct approach, through the use of iterative stationary and Krylov subspace methods up to the combination of direct and iterative techniques including preconditioning. Significantly less attention however has been paid so far to the numerical behavior of saddle-point solvers. In this section we concentrate on the numerical behavior of the Schur complement reduction method and the null-space method which compute the sequences of approximate solutions x_{k+1} and y_{k+1} for $k = 0, 1, \ldots$: one of them is first obtained in the outer iteration from a reduced system of a smaller dimension, and once it has been computed, the other approximate solution is obtained by back-substitution solving another reduced problem in the inner iteration loop. The computation of such approximations can be very expensive and time consuming. Therefore, in practice some relaxations are necessary in various stages of computation, and the solution of certain subproblems intentionally approximated with some inexact process very often represented by some iterative method that is terminated with some prescribed tolerance level. In addition, the effects of rounding errors in finite precision arithmetic have to be also taken into account, and they may significantly contribute to the behavior of the saddle-point solver. Therefore, we have to be aware that there are certain limitations of such relaxation strategies. Indeed, one must expect that inexact computations can lead to convergence delay and to limitations on the maximum attainable accuracy of computed approximate solutions (see the schematic Fig. 8.1).

Numerous inexact schemes have been used and analyzed (see, e.g., the analysis of inexact Uzawa's algorithms and inexact null-space methods). These works contain mainly the analysis of a convergence delay caused by the inexact solution of inner systems or least squares problems. In this chapter we concentrate on the question of what is the best accuracy we can get from the inexact Schur complement method or from the inexact null-space method. Since a complete analysis of all sources of inexactness is beyond the scope of this contribution, without going into

© Springer Nature Switzerland AG 2018
M. Rozložník, *Saddle-Point Problems and Their Iterative Solution*,
Nečas Center Series, https://doi.org/10.1007/978-3-030-01431-5_8

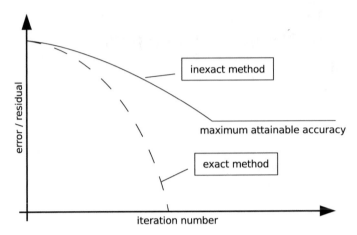

Fig. 8.1 Delay of convergence and the maximum attainable accuracy

details and without rigorous proofs of main results, we discuss only the following three illustrative cases:

- the influence of inexact solution of inner systems on the maximum attainable accuracy in the outer iteration of the Schur complement method or of the null-space method, addressing also the question whether large intermediate approximate solutions reduce the accuracy of this inner-outer iteration schemes for nonsymmetric saddle-point systems;
- the influence of inexact solution of inner systems on the maximum attainable accuracy of approximate solutions in the Schur complement or in the null-space projection method with respect to the back-substitution formula used for computing the other approximate solution to saddle-point systems (we point out the optimal implementations delivering high-accuracy approximate solutions);
- the influence of the scaling of the diagonal block in the saddle-point system solved with the conjugate gradient method and preconditioned with the constraint preconditioner on the accuracy approximate solutions computed by these variants of the Schur complement method or of null-space method.

Throughout this chapter, we consider the saddle-point problem (1.1) with (1.3) and (1.4), where A is symmetric positive definite and B has a full-column rank. Numerical experiments are performed on the small model example taken from [71]: we set $m = 100$, $n = 20$ and use the Matlab notation to define

$$A = \text{tridiag}(1, 4, 1) \in \mathcal{R}^{m,m}, B = \text{rand}(m, n) \in \mathcal{R}^{m,n},$$

$$c = \text{rand}(m, 1) \in \mathcal{R}^{m}. \tag{8.1}$$

The spectrum of A and the singular value set of B lies in the interval $[2.0010, 5.9990]$ and $[2.1727, 7.1695]$, respectively. Therefore, the conditioning of A or B does not play an important role in our experiments. We use the standard double precision arithmetic with the roundoff unit $u = 1.11e - 16$.

8.1 Accuracy of Approximate Solutions in Outer Iteration

In this subsection we look at the influence of inexact solution of inner systems on the maximum attainable accuracy of approximate solutions computed in the outer iteration of the Schur complement and null-space methods.

Schur complement method Because the solution of systems with the matrix block A in the Schur complement method (see Algorithm 4.1 in Chap. 4) can be expensive, these systems are in practice solved only approximately. Our analysis is based on the assumption that every solution of a system with A is replaced by an approximate solution produced by some suitably chosen solver. The resulting vector is then interpreted as an exact solution of the system with the same right-hand side vector but with a perturbed matrix $A + \Delta A$. We always require that the relative norm of the perturbation is bounded as $\|\Delta A\| \leq \tau \|A\|$, where τ measures the accuracy of the computed approximate solution. We will also assume that the perturbation ΔA does not exceed the limitation given by the distance of A to the nearest singular matrix and put a restriction in the form $\tau \kappa(A) \ll 1$. Note that if $\tau = O(u)$, then we have a backward stable method [40]. In our numerical experiments, we solve the systems with A inexactly using either some suitable iterative method or using a backward stable direct method as will be indicated by the notation $\tau = O(u)$. If A is symmetric positive definite, then we use the conjugate gradient method or the direct method using the Cholesky factorization [40].

For distinction, we will denote the quantities computed in finite precision arithmetic by bars. It was shown in Theorem 2.1 of [45] that the gap between the true residual $-B^T A^{-1} c + B^T A^{-1} B \bar{y}_{k+1}$ and the updated residual $\bar{r}_{k+1}$ can be bounded as

$$\| - B^T A^{-1} c + B^T A^{-1} B \bar{y}_{k+1} - \bar{r}_{k+1}\| \leq \frac{O(\tau)\kappa(A)}{1 - \tau\kappa(A)} \|A^{-1}\| \|B\| (\|c\| + \|B\| \bar{Y}_{k+1}),$$

$$(8.2)$$

where $\bar{Y}_{k+1}$ is defined as a maximum over all norms of previously computed approximate solutions $\bar{Y}_{k+1} = \max_{i=0,1,\dots,k+1} \|\bar{y}_i\|$.

It is a well-known fact that the residual $\bar{r}_{k+1}$ computed recursively via (4.9) usually converges far below $O(u)$. Using this assumption, we can obtain from the estimate for the gap $-B^T A^{-1} c + B^T A^{-1} B \bar{y}_{k+1} - \bar{r}_{k+1}$ the estimate for the maximum attainable accuracy of the true residual $-B^T A^{-1} c + B^T A^{-1} B \bar{y}_{k+1}$ itself. Summarizing, while the updated residual $\bar{r}_{k+1}$ converges to zero, the true residual stagnates at the level proportional to τ. This is also illustrated in our numerical example (8.1), where the Schur complement system (4.3) is solved using

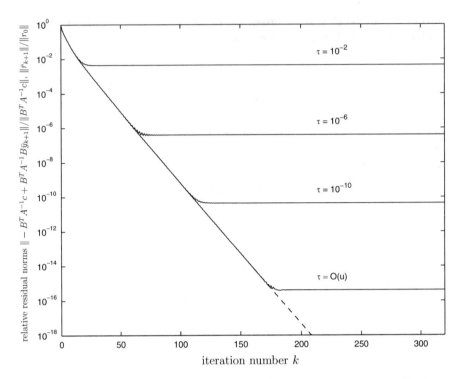

Fig. 8.2 Schur complement method: the relative norms of the true residual $-B^T A^{-1} c + B^T A^{-1} B \bar{y}_{k+1}$ (solid lines) and the updated residual $\bar{r}_{k+1}$ (dashed lines)

the conjugate gradient method with the initial approximation y_0 set to zero. In Fig. 8.2 we show the relative norms of the true residual $-B^T A^{-1} c + B^T A^{-1} B \bar{y}_{k+1}$ (solid lines) and the updated residual $\bar{r}_{k+1}$ (dashed lines) for several values of the parameter τ.

Similarly to Greenbaum [37] we have shown that the gap between the true and updated residual is proportional to the maximum norm of approximate solutions computed during the whole iteration process. Since the Schur complement system is symmetric negative definite, the norm of the error or residual converges monotonically for the most iterative methods like the steepest descent, the conjugate gradient, conjugate residual method, or other error/residual minimizing methods; or at least becomes orders of magnitude smaller than initial error/residual without exceeding this limit. In such cases, the quantity $\bar{y}_{k+1}$ does not play an important role in the bound, and it can be usually replaced by $\|y_0\|$ or a small multiple of $\|y\|$. The situation is more complicated when A is nonsingular and nonsymmetric; see the discussion in the end of this subsection or [46].

Null-space method In contrast to the Schur complement method, the inexactness in the null-space method is connected with the matrix B instead of A. The least squares with the matrix B in Algorithm 4.2 are in practice solved inexactly. In our considerations we assume that each computed solution of the least squares problem with B is seen as an exact solution of a perturbed problem with the matrix $B + \Delta B$, where $\|\Delta B\|/\|B\| \leq \tau$. The parameter τ again represents the measure for inexact solution of any least squares problem with B, and it describes the backward error. This can be achieved in various ways considering the inner iteration loop solving the associated system of normal equations (2.2), the augmented system formulation (2.5) or using a backward stable direct method for least squares problem (2.1). We assume $\tau\kappa(B) \ll 1$ which guarantees $B + \Delta B$ to have a full-column rank. Note that if $\tau = O(u)$, we have a backward stable method for solving the least squares problem with B. In our experiments we applied the conjugate gradient method on (2.2) with the stopping criterion based on the corresponding backward error. Notation $\tau = O(u)$ stands for the backward stable solution via Householder QR factorization [40].

Theorem 3.1 in [45] shows that the true residual $ZZ^T(c - AZZ^T\bar{x}_{k+1})$ of the approximate solution $\bar{x}_{k+1}$ computed in the null-space method is proportional to τ, provided that the updated residual $\bar{r}_{k+1}$ converges far below the level of unit roundoff. Indeed, the gap between the true residual $ZZ^T(c - AZZ^T\bar{x}_{k+1})$ and the projection of the updated residual $ZZ^T\bar{r}_{k+1}$ can be bounded by

$$\|ZZ^T(c - AZZ^T\bar{x}_{k+1} - \bar{r}_{k+1})\| \leq \frac{O(\tau)\kappa(B)}{1 - \tau\kappa(B)}(\|c\| + \|A\|\bar{X}_{k+1}), \qquad (8.3)$$

where $\bar{X}_{k+1} = \max_{i=0,1,\ldots,k+1}\|\bar{x}_i\|$. In Fig. 8.3 we report the relative norms of the true residual $ZZ^T(c - AZZ^T\bar{x}_{k+1})$ (solid lines) and the updated residual $\bar{r}_{k+1}$ (dashed lines) on the same example (8.1) for several values of the parameter τ. The numerical results confirm that the true residual $Z^T(c - AZZ^T\bar{x}_{k+1})$ is well approximated by $Z^T\bar{r}_{k+1}$, until they reach the level $O(\tau)$ when $\bar{r}_{k+1}$ continues to decrease and $ZZ^T(c - AZZ^T\bar{x}_{k+1})$ stagnates on the level of its maximum attainable accuracy.

Nonsymmetric saddle-point problems As it was already noted, the bounds developed for the residual gap in the Schur complement and null-space methods depend on the largest norm of computed approximate solutions (either $\bar{x}_i$ or $\bar{y}_i$) during the whole iteration process $i = 0, 1, \ldots, k + 1$. If A is symmetric positive definite, the matrices $B^T A^{-1} B$ and $Z^T AZ$ are also symmetric positive definite, and the CG method is used for solving the systems (4.3) and (4.14). The error of approximate solutions in the CG method is known to converge monotonously in the energy norm, and thus the factors $\bar{X}_{k+1}$ and $\bar{Y}_{k+1}$ in (8.3) and (8.2), respectively, are comparable to the norms of exact solutions x_* and y_*. In such cases, these terms do not play an important role.

However, if the matrix block A is nonsymmetric, the matrices $B^T A^{-1} B$ and $Z^T AZ$ are nonsymmetric, and some nonsymmetric Krylov subspace must be used for solving the systems (4.3) and (4.14). Since the GMRES method with approx-imate solutions satisfying the residual minimization property (6.31) is expensive, Krylov subspace methods with lower and roughly constant work and storage

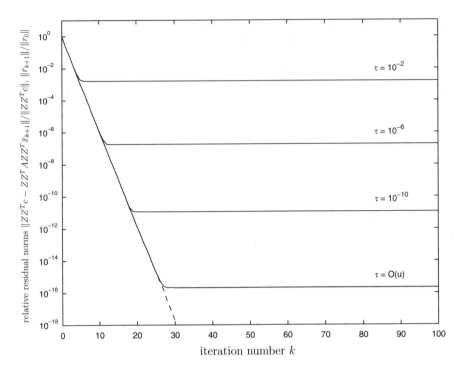

Fig. 8.3 Null-space method: the relative norms of the true residual $Z^T(c - AZZ^T\bar{x}_{k+1})$ (solid lines) and the updated residual $\bar{r}_{k+1}$ (dashed lines)

requirements per step are used. One of the most straightforward methods to solve a general nonsymmetric method is the CGNE method that is just the CG method applied to the symmetric positive definite system of related normal equations. Its approximate solutions minimize the error norm over certain Krylov subspace method, but its convergence is very often too slow. On the other hand, other non-optimal Krylov subspace methods with short recurrences such as Bi-CG, QMR, or CGS are known to produce very large intermediate approximate solutions. The oscillation of their norms may then affect the maximum attainable accuracy of these schemes [37].

In the following we show the behavior of various Krylov subspace methods chosen for solving the Schur complement system (4.3) and the null-space projected system (4.3). In particular, we consider GMRES, Bi-CG, CGS, and CGNE on the small nonsymmetric example with $m = 100$ and $n = 50$,

$$A = \text{tridiag}(1, 10^{-5}, -1) \in \mathcal{R}^{m,m}, \, B = \text{rand}(m, n) \in \mathcal{R}^{m,n},$$

$$c = \text{ones}(m, 1) \in \mathcal{R}^m. \tag{8.4}$$

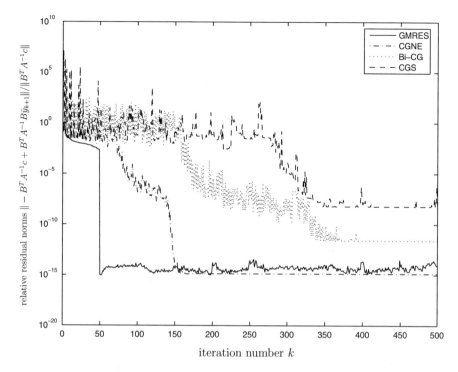

Fig. 8.4 The relative norms of the residual $-B^T A^{-1} c + B^T A^{-1} B \bar{y}_{k+1}$ in the Schur complement method with respect to iteration number for GMRES (solid lines), CGNE (dash-dotted lines), BiCG (dotted lines), and CGS (dashed lines)

Note that these matrices are well-conditioned as $\kappa(A) = \|A\| \|A^{-1}\| \approx 2.00 \cdot 32.15 \approx 64.30$ and $\kappa(B) = \|B\|/\sigma_{\min}(B) \approx 7.39 \cdot 0.75 \approx 5.55$. For each test we take the initial guess $y_0 = 0$ or $x_0 = 0$. The inner systems with the matrix A in the Schur complement method are solved with the direct method based on the LU factorization of A, and the inner least squares problems with B in the null-space method are solved using the Householder QR factorization.

In Fig. 8.4 we show the true residual $-B^T A^{-1} c + B^T A^{-1} B \bar{y}_{k+1}$ in the Schur complement method for GMRES (solid lines), CGNE (dash-dotted lines), Bi-CG (dotted lines), and CGS (dashed lines). Similarly, in Fig. 8.5 we report the relative norms of the residual $ZZ^T(c - AZZ^T \bar{x}_{k+1})$ in the null-space method for GMRES (solid lines), CGNE (dash-dotted lines), BiCG (dotted lines), and CGS (dashed lines). It is clear from Figs. 8.4 and 8.5 that for error norm minimizing CGNE and the residual minimizing GMRES is the maximum attainable accuracy level proportional to the roundoff unit. The quantities $\bar{Y}_{k+1}$ and $\bar{X}_{k+1}$ are comparable to the norms of y_* and x_*, and they do not significantly affect the limiting accuracy of computed approximate solutions. The situation is completely different for Bi-CG and CGS, where the norm of approximate solutions grows up to 10^5 (for Bi-CG) and to 10^7 (for CGS) in the Schur complement method, and to 10^6 (for Bi-CG) and

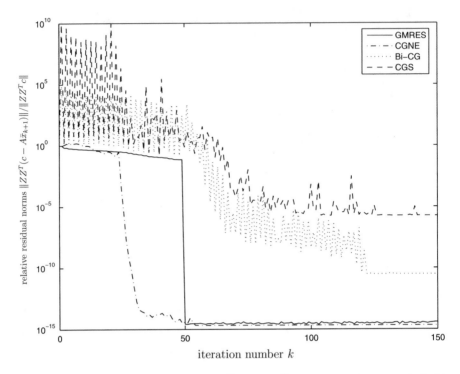

Fig. 8.5 The relative norms of the residual $ZZ^T(c - AZZ^T\bar{x}_{k+1})$ in the null-space method with respect to iteration number for GMRES (solid lines), CGNE (dash-dotted lines), BiCG (dotted lines), and CGS (dashed lines)

Table 8.1 The quantities		Schur complement method	Null-space method
$\bar{Y}_{k+1}$ and $\bar{X}_{k+1}$ in the Schur		$\bar{Y}_{k+1}$	$\bar{X}_{k+1}$
complement method and in the null-space method for GMRES, CGNE, BiCG, and CGS	GMRES	1.6155 10^1	3.9445 10^1
	CGNE	1.6157 10^1	3.9445 10^1
	BiCG	9.8556 10^4	6.5733 10^5
	CGS	3.3247 10^7	5.2896 10^{10}

to 10^{11} (for CGS) in the null-space method (see corresponding Table 8.1). Indeed, the results confirm that the residuals reach ultimately the levels which are roughly $O(u)\bar{Y}_{k+1}$ or $O(u)\bar{X}_{k+1}$ instead of $O(u)$. Note that the matrices A and B are well-conditioned and thus the conditioning of the matrices $B^T A^{-1} B$, and $Z^T AZ$ does not affect the maximum attainable accuracy level on these examples.

8.2   Back-Substitution and Accuracy of Inexact Saddle-Point Solvers

In this subchapter we illustrate that the choice of the back-substitution formula in the Schur complement method or in the null-space method can considerably affect the maximum attainable accuracy of approximate solutions computed in finite precision arithmetic. Here we concentrate on the question what is the best accuracy we can get from schemes, where the inner systems or the inner least square problems are solved inexactly. As it will be seen later, it significantly depends on the back-substitution formula used for computing the remaining unknowns.

Schur complement method   We will distinguish between three mathematically equivalent back-substitution formulas

$$x_{k+1} = x_k + \alpha_k(-A^{-1}Bq_k), \tag{8.5}$$

$$x_{k+1} = A^{-1}(c - By_{k+1}), \tag{8.6}$$

$$x_{k+1} = x_k + A^{-1}(c - Ax_k - By_{k+1}). \tag{8.7}$$

The resulting schemes are summarized in Algorithm 8.1. They are used in many applications, including classical Uzawa-type algorithms, two-level pressure correction approach, or in the context of inner-outer iteration methods for solving the saddle-point problem (1.1) with (1.3) and (1.4). The inner systems with the matrix A are assumed to be solved inexactly, and we use the same backward approach as in the previous subsection, i.e., each approximate solution to a system with A is interpreted as an exact solution of a perturbed system with a relative perturbation determined by the parameter τ. To preserve the nonsingularity of perturbed systems, we assume that $\tau\kappa(A) \ll 1$.

The schemes (8.5), (8.6), and (8.7) were analyzed in detail, and the bounds on the maximum attainable accuracy for the two components $c - A\bar{x}_{k+1} - B\bar{y}_{k+1}$ and $-B^T\bar{x}_{k+1}$ of the true residual $\mathbb{b} - \mathbb{A}\bar{\mathbb{x}}_{k+1}$ were given in [45]. Indeed, it was shown in Theorems 2.2, 2.3, and 2.4 that the true residual $c - A\bar{x}_{k+1} - B\bar{y}_{k+1}$ and the gap between the residuals $-B^T\bar{x}_{k+1}$ and $\bar{r}_{k+1}$ are bounded by

$$\|c - A\bar{x}_{k+1} - B\bar{y}_{k+1}\| \le \frac{O(\alpha_1)\kappa(A)}{1 - \tau\kappa(A)}(\|c\| + \|B\|\bar{Y}_{k+1}), \tag{8.8}$$

$$\| - B^T\bar{x}_{k+1} - \bar{r}_{k+1}\| \le \frac{O(\alpha_2)\kappa(A)}{1 - \tau\kappa(A)}\|A^{-1}\|\|B\|(\|c\| + \|B\|\bar{Y}_{k+1}), \tag{8.9}$$

Algorithm 8.1 Schur complement method with back-substitution formulas

---

choose y_0, solve $Ax = c - By_0$ for x_0

for $k = 0, 1, \ldots$

compute α_k and q_k

$y_{k+1} = y_k + \alpha_k q_k$

  solve $Ap_k = -Bq_k$

  back-substitution:

  A: $x_{k+1} = x_k + \alpha_k p_k,$

  B: solve $Ax_{k+1} = c - By_{k+1}$

  C: solve $Au_k = c - Ax_k - By_{k+1},\ x_{k+1} = x_k + u_k.$

$r_{k+1} = r_k - \alpha_k B^T p_k$

end

Table 8.2 The factors α_1 and α_2 for the schemes (8.5), (8.6), and (8.7) in the Schur complement method

Back-substitution scheme	α_1	α_2
A: Updated approximate solution (8.5)	τ	u
$x_{k+1} = x_k + \alpha_k p_k$		
B: Direct substitution (8.6)	τ	τ
$x_{k+1} = A^{-1}(c - By_{k+1})$		
C: Corrected substitution (8.7)	u	τ
$x_{k+1} = x_k + A^{-1}(c - Ax_k - By_{k+1})$		

where $\bar{Y}_{k+1} = \max_{i=0,1,\ldots,k+1} \|\bar{y}_i\|$. The factors α_1 and α_2 depend on the choice of the back substitution formula (8.5), (8.6) or (8.7), and they are equal to the values given in Table 8.2.

   Provided that the updated residual $\bar{r}_{k+1}$ converges below the level of the roundoff unit u, the use of the scheme with the updated approximate solution (8.5) thus leads to the second component of the residual $-B^T \bar{x}_{k+1}$ on the level proportional to u. However, the first component of the residual $c - A\bar{x}_{k+1} - B\bar{y}_{k+1}$ is proportional to the parameter τ. When using the direct substitution scheme (8.6), both residuals stagnate ultimately on the level proportional to the parameter τ. The corrected substitution scheme (8.7) gives a similar accuracy in the second component $-B^T \bar{x}_{k+1}$, but it is significantly more accurate in the first component $c - A\bar{x}_{k+1} - B\bar{y}_{k+1}$ that is ultimately proportional to u, independent of the fact that all inner systems are solved inexactly with the accuracy proportional to τ. Note that in practical situations, the parameter τ is significantly larger than the roundoff unit u. These results essentially show that, depending on which back-substitution formula is used, the approximate solutions computed by such inexact saddle-point

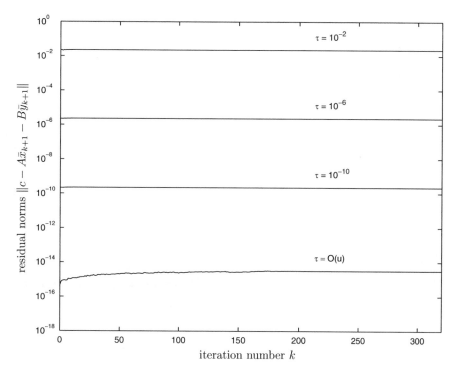

Fig. 8.6 Schur complement method: the norms of the true residual $c - A\bar{x}_{k+1} - B\bar{y}_{k+1}$ (solid lines) for the updated approximate solution scheme (8.5)

solvers may satisfy either the first or the second block equation of (1.1) with (1.3) and (1.4) to the working accuracy.

In our numerical illustrations, we consider again the simple model problem (8.1). For solving the inner systems with A, we use the conjugate gradient method that terminates when it reaches the approximate solution with the backward error (the definition can be found in [40]) smaller or equal to the parameter τ or the backward stable Cholesky factorization that corresponds to $\tau = O(u)$. Therefore, the factor $\bar{Y}_{k+1}$ in (8.8) and (8.9) is not significant here. Figures 8.6 and 8.7 show the norms of the true residual $c - A\bar{x}_{k+1} - B\bar{y}_{k+1}$ and $-B^T \bar{x}_{k+1}$ (solid lines) together with the norms of the updated residuals $\bar{r}_{k+1}$ (dashed lines) for the scheme (8.5). Figure 8.8 illustrates the norms of $-B^T \bar{x}_{k+1}$ (solid lines) and $\bar{r}_{k+1}$ (dashed lines) for the scheme (8.6). Figure 8.9 shows the norms of the true residual $c - A\bar{x}_{k+1} - B\bar{y}_{k+1}$ for the scheme (8.7). All numerical results are in good agreement with the bounds (8.8) and (8.9) specified in Table 8.2.

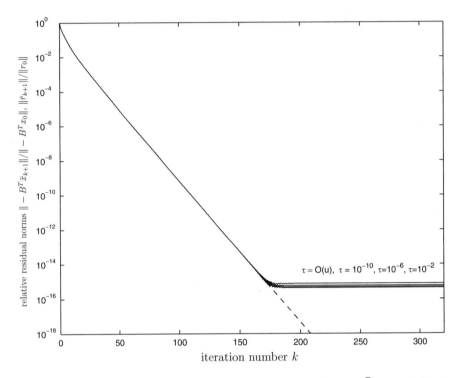

Fig. 8.7 Schur complement method: the relative norms of the true residual $-B^T \bar{x}_{k+1}$ (solid lines) and the updated residual $\bar{r}_{k+1}$ (dashed lines) for the updated approximate solution scheme (8.5)

Null-space method We distinguish between three mathematically equivalent back-substitution formulas

$$y_{k+1} = y_k + q_k, \quad q_k = \arg \min_{y \in \mathcal{R}^n} \|(r_k - \alpha_k A p_k) - By\|, \qquad (8.10)$$

$$y_{k+1} = \arg \min_{y \in \mathcal{R}^n} \|(c - Ax_{k+1}) - By\|, \qquad (8.11)$$

$$y_{k+1} = y_k + v_k, \quad v_k = \arg \min_{y \in \mathcal{R}^n} \|(c - Ax_{k+1} - By_k) - By\|. \qquad (8.12)$$

The resulting schemes are summarized in Algorithm 8.2. They are often used for solving quadratic programming problems, multigrid methods, or constraint preconditioning. The least squares problems with the matrix B are in practice solved inexactly. Each computed approximate solution of the least squares problem is interpreted as an exact solution of a perturbed least squares problem with the relative perturbation that is smaller or equal to the parameter τ. The measure for the inexact solution of the inner least squares problems is thus again represented by the backward error [40]. We assume that $\tau \kappa(B) \ll 1$ which guarantees that each

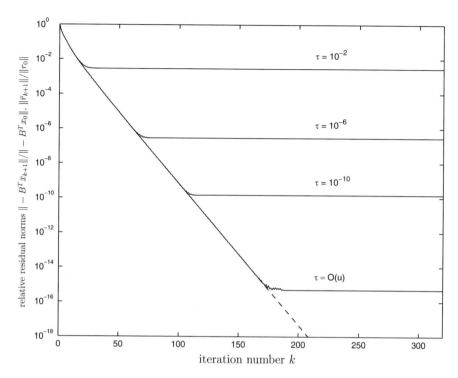

Fig. 8.8 Schur complement method: the relative norms of the true residual $-B^T \bar{x}_{k+1}$ (solid lines) and the updated residual $\bar{r}_{k+1}$ (dashed lines) for the direct substitution scheme (8.6)

perturbed problem has a full-column rank. Note that if $\tau = O(u)$, then we have a backward stable method for solving the least squares problem with B.

The schemes (8.10), (8.11), and (8.12) were analyzed in Theorems 3.2, 3.3, and 3.4 of [45]. It was proved that the gap between the residuals $c - A\bar{x}_{k+1} - B\bar{y}_{k+1}$ and $\bar{r}_{k+1}$ and the true residual $-B^T \bar{x}_{k+1}$ satisfy the bounds

$$\|c - A\bar{x}_{k+1} - B\bar{y}_{k+1} - \bar{r}_{k+1}\| \leq \frac{O(\alpha_3)\kappa(B)}{1 - \tau\kappa(B)}(\|c\| + \|A\|\bar{X}_{k+1}), \quad (8.13)$$

$$\| - B^T \bar{x}_{k+1}\| \leq \frac{O(\tau)\kappa(B)}{1 - \tau\kappa(B)}\|B\|\bar{X}_{k+1}, \quad (8.14)$$

where $\bar{X}_{k+1} = \max\{\|\bar{x}_i\| \mid i = 0, 1, \ldots, k+1\}$. The factor α_3 depends on the choice of the back substitution formula (8.10), (8.11), or (8.12), and it is equal to the values given in Table 8.2.

The residual $-B^T \bar{x}_{k+1}$ obviously does not depend on the back-substitution scheme, but it does depend on the recurrences (4.18) and (4.19) used in the outer iteration of the null-space method. Since the projection onto $\mathcal{N}(B^T)$ in (4.18) and (4.19) is applied inexactly, it is quite natural that the second component of

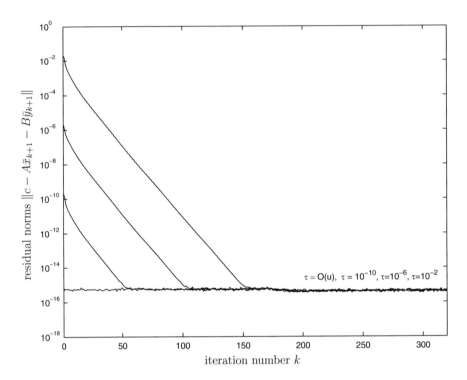

Fig. 8.9 Schur complement method: the norms of the true residual $c - A\bar{x}_{k+1} - B\bar{y}_{k+1}$ (solid lines) for the corrected substitution scheme (8.7)

the residual $-B^T\bar{x}_{k+1}$ is proportional to the parameter τ. Provided that the updated residual $\bar{r}_{k+1}$ converges below the level of the roundoff unit u, the use of the updated approximate solution scheme (8.10) and the use of the corrected substitution scheme (8.12) lead to the first component of the residual $c - A\bar{x}_{k+1} - B\bar{y}_{k+1}$ on the level proportional to u, independent of the fact that all inner systems are solved inexactly with the accuracy proportional to τ. However, when using the direct substitution scheme (8.11), the first component of the residual $c - A\bar{x}_{k+1} - B\bar{y}_{k+1}$ is proportional to the parameter τ (Table 8.3). Note again that in practice, the parameter τ is significantly larger than the roundoff unit u.

In our experiments, we consider the same saddle-point problem as for the Schur complement method. For solving the inner least squares problems, we use the conjugate gradient method applied to the system of normal equations (2.2) with the stopping criterion based on the backward error and with the threshold τ. Therefore, the factor $\bar{X}_{k+1}$ in (8.13) and (8.14) is comparable to the size

---

Algorithm 8.2 Null-space method with back-substitution formulas

---

choose x_0, solve $\min\limits_{y \in \mathcal{R}^n} \|(c - Ax_0) - By\|$ for y_0

for $k = 0, 1, \ldots$

compute α_k and $p_k \in N(B^T)$

$x_{k+1} = x_k + \alpha_k p_k$

  back-substitution:

  A: $y_{k+1} = y_k + q_k$, solve $q_k = \arg\min\limits_{y \in \mathcal{R}^n} \|(r_k - \alpha_k Ap_k) - By\|$

  B: solve $y_{k+1} = \arg\min\limits_{y \in \mathcal{R}^n} \|(c - Ax_{k+1}) - By\|$

  C: solve $v_k = \arg\min\limits_{y \in \mathcal{R}^n} \|(c - Ax_{k+1} - By_k) - By\|$,

    $y_{k+1} = y_k + v_k.$

$\tilde{r}_{k+1} = \tilde{r}_k - \alpha_k Z^T Ap_k$

$r_{k+1} = Z\tilde{r}_{k+1}$

end

Table 8.3 The factor α_3 for the schemes (8.10), (8.11), and (8.12) in the null-space method

Back-substitution scheme	α_3
A: Updated approximate solution (8.10)	u
$y_{k+1} = y_k + \arg\min_{y \in \mathcal{R}^n} \|(r_k - \alpha_k Ap_k) - By\|$	
B: Direct substitution (8.11)	τ
$y_{k+1} = \arg\min_{y \in \mathcal{R}^n} \|(c - Ax_{k+1}) - By\|$	
C: Corrected substitution (8.12)	u
$y_{k+1} = y_k + \arg\min_{y \in \mathcal{R}^n} \|(c - Ax_{k+1} - By_k) - By\|$	

of the exact solution x_*, and it does not play a significant role here. The case $\tau = O(u)$ corresponds to the least squares problem solved via the backward stable Householder QR factorization [40]. In Fig. 8.10 we show the norms of the true residual $c - A\bar{x}_{k+1} - B\bar{y}_{k+1}$ (solid lines) together with the norms of the updated residuals $\bar{r}_{k+1}$ (dashed lines) for the scheme (8.10). The same quantities are reported in Fig. 8.11 for the direct substitution scheme (8.11) and in Fig. 8.12 for the corrected substitution scheme (8.12). Figure 8.13 shows the true residual $-B^T \bar{x}_{k+1}$ that does not depend on the back-substitution formula. Our numerical results are in a good agreement with the bounds (8.13) and (8.14).

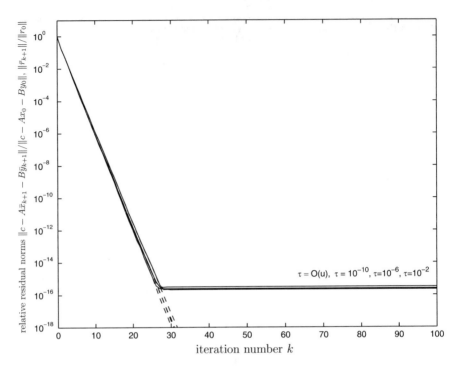

Fig. 8.10 Null-space method: the relative norms of the true residual $c - A\bar{x}_{k+1} - B\bar{y}_{k+1}$ (solid lines) and the updated residual $\bar{r}_{k+1}$ (dashed lines) for the updated approximate solution scheme (8.10)

8.3 Conjugate Gradient Method with Constraint Preconditioning

In this subsection we consider the saddle-point problem (1.1) with (1.3) and (1.4), where A is symmetric positive definite and B has a full-column rank, and we study the numerical behavior of the CG method applied to (1.1) and preconditioned with the constraint preconditioners (7.21) and (7.31).

We show that although the indefiniteness of the system matrix and the preconditioner in some cases do not make the algorithm robust and its breakdown may occur, a suitable scaling of the problem or other safeguarding procedures can be used to avoid its misconvergence and to improve the accuracy of approximate solutions in finite precision arithmetic.

Schur complement method First, we consider the constraint preconditioner (7.21) to solve the preconditioned system (7.27). As we have shown in the previous subsection, if the initial guess $\mathbb{x}_0$ is chosen so that $\mathbb{r}_0 = \begin{pmatrix} 0 \\ r_0 \end{pmatrix}$, then the residuals of

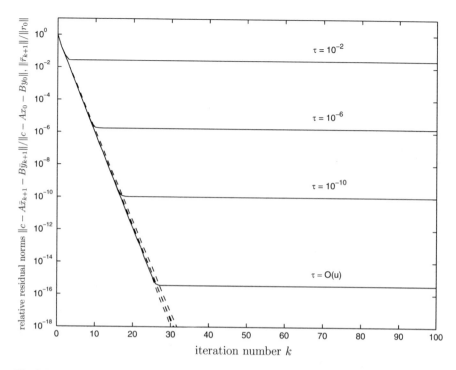

Fig. 8.11 Null-space method: the relative norms of the true residual $c - A\bar{x}_{k+1} - B\bar{y}_{k+1}$ (solid lines) and the updated residual $\bar{r}_{k+1}$ (dashed lines) for the direct substitution scheme (8.11)

any Krylov method have a special form $\mathbb{r}_{k+1} = \begin{pmatrix} 0 \\ r_{k+1} \end{pmatrix}$. In the preconditioned CG method, the residual $\mathbb{r}_{k+1}$ is orthogonal to the subspace $\mathbb{P}^{-1}\mathcal{K}_{k+1}(\mathbb{A}\mathbb{P}^{-1}, \mathbb{r}_0)$ as

$$\mathbb{r}_{k+1} \in \mathbb{r}_0 + \mathbb{A}\mathbb{P}^{-1}\mathcal{K}_{k+1}(\mathbb{A}\mathbb{P}^{-1}, \mathbb{r}_0), \quad \mathbb{r}_{k+1} \perp \mathbb{P}^{-1}\mathcal{K}_{k+1}(\mathbb{A}\mathbb{P}^{-1}, \mathbb{r}_0). \quad (8.15)$$

Indeed, (8.15) shows that $\mathbb{r}_{k+1}$ is orthogonal to all previous residuals $\mathbb{r}_j$, $j = 0, \ldots, k$, with respect to the bilinear form induced by the symmetric indefinite matrix $\mathbb{P}^{-1}$ satisfying

$$\mathbb{r}_{k+1}^T \mathbb{P}^{-1} \mathbb{r}_j = 0, \quad j = 0, \ldots, k. \quad (8.16)$$

Substituting for the residuals $\mathbb{r}_j = \begin{pmatrix} 0 \\ r_j \end{pmatrix}$, $j = 0, \ldots, k + 1$, and for $\mathbb{P}^{-1}$ the identity (7.23), we get that $r_{k+1}^T r_j = 0$ for $j = 0, \ldots, k$, or equivalently, $r_{k+1} \perp \mathcal{K}_k(B^T A^{-1} B, r_0)$. Since we have also $A x_{k+1} + B y_{k+1} = c$, the vector $r_{k+1} = -B^T x_{k+1} = -B^T A^{-1} c + B^T A^{-1} B y_{k+1}$ can be seen as the residual vector of the approximate solution y_{k+1} from the unpreconditioned CG method applied to the

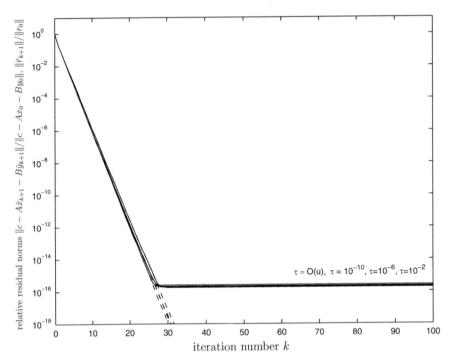

Fig. 8.12 Null-space method: the relative norms of the true residual $c - A\bar{x}_{k+1} - B\bar{y}_{k+1}$ (solid lines) and the updated residual $\bar{r}_{k+1}$ (dashed lines) for the updated approximate solution scheme (8.12)

symmetric negative definite Schur complement system $-B^T A^{-1} By = -B^T A^{-1}c$. Therefore, it satisfies the $B^T A^{-1} B$-energy error norm minimization property

$$\|y_* - y_{k+1}\|_{B^T A^{-1}B} = \min_{y \in y_0 + K_{k+1}(B^T A^{-1}B, B^T A^{-1}c-d)} \|y_* - y\|_{B^T A^{-1}B}. \qquad (8.17)$$

Thus, we get the Schur complement method, where the outer iteration is solved by the CG method (see Algorithm 4.1). This algorithm is summarized in Algorithm 8.3.

Null-space method We can also consider the CG method applied to the preconditioned system (7.27) with the constraint preconditioner (7.31). The scheme of the method is summarized in Algorithm 8.4. Indeed, if the initial guess x_0 is chosen so that $r_0 = \begin{pmatrix} r_0 \\ 0 \end{pmatrix}$, then the residuals of any Krylov method with this preconditioner have a special form $r_{k+1} = \begin{pmatrix} r_{k+1} \\ 0 \end{pmatrix}$. We will use the same argument as for the Schur complement method. The residual r_{k+1} satisfies (8.15), and due to (8.16), it is mutually orthogonal with respect to the bilinear form induced by the matrix

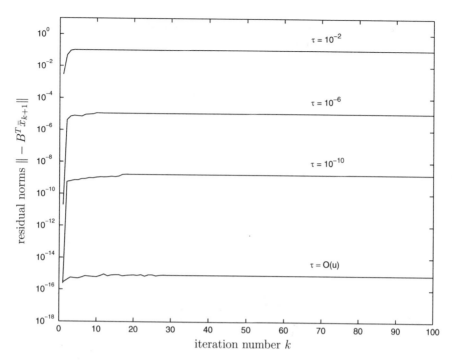

Fig. 8.13 Null-space method: the norms of the true residual $-B^T \bar{x}_{k+1}$ (solid lines) for the updated approximate solution scheme (8.10)

$\mathbb{P}^{-1}$. Using (7.33), and (7.35), and $\mathbb{r}_{k+1} = \mathbb{A}\mathbb{e}_{k+1}$, the error $\mathbb{e}_{k+1} = \begin{pmatrix} x_* - x_{k+1} \\ y_* - y_{k+1} \end{pmatrix}$ satisfies the conditions

$$\mathbb{e}_{k+1}^T \mathbb{A}\mathbb{P}^{-1}\mathbb{r}_j = (x_* - x_{k+1})^T (AZZ^T + B(B^T B)^{-1} B^T)r_j = 0 \qquad (8.18)$$

for $j = 0, \ldots, k$. Since $x_* - x_{k+1}$ belongs to $\mathcal{N}(B^T)$, we immediately see that $(x_* - x_{k+1})^T ZZ^T AZZ^T r_j = 0$. Consequently, the vector x_{k+1} can be seen as an approximate solution from the CG method applied to the symmetric positive semidefinite system

$$ZZ^T AZZ^T x = ZZ^T c \qquad (8.19)$$

satisfying the A-energy error norm minimization property

$$\|x_* - x_{k+1}\|_A = \min_{x \in x_0 + \mathrm{span}\{ZZ^T r_0, \ldots, ZZ^T r_k\}} \|x_* - x\|_A . \qquad (8.20)$$

Thus we get the null-space projection method (see Algorithm 4.2) with outer iteration solved with the conjugate gradient method, where the vector x_{k+1} converges to

---

Algorithm 8.3 Preconditioned CG with the constraint preconditioner (7.21) in the Schur complement method

---

$$\text{choose } \begin{pmatrix} x_0 \\ y_0 \end{pmatrix}, \quad \mathbb{r}_0 = \mathbb{b} - \mathbb{A} \begin{pmatrix} x_0 \\ y_0 \end{pmatrix} = \begin{pmatrix} 0 \\ r_0 \end{pmatrix}$$

$$\begin{pmatrix} p_0 \\ q_0 \end{pmatrix} = \mathbb{P}^{-1}\mathbb{r}_0 = \mathbb{P}^{-1}\begin{pmatrix} 0 \\ r_0 \end{pmatrix}$$

for $k = 0, 1, \dots$

$$\alpha_k = \frac{\left(\begin{pmatrix} 0 \\ r_k \end{pmatrix}, \mathbb{P}^{-1}\begin{pmatrix} 0 \\ r_k \end{pmatrix} \right)}{\left(\mathbb{A}\begin{pmatrix} p_k \\ q_k \end{pmatrix}, \begin{pmatrix} p_k \\ q_k \end{pmatrix} \right)} \qquad\qquad \alpha_k = \frac{(\mathbb{r}_k, \mathbb{P}^{-1}\mathbb{r}_k)}{(\mathbb{A}\mathbb{p}_k, \mathbb{p}_k)}$$

$$\begin{pmatrix} x_{k+1} \\ y_{k+1} \end{pmatrix} = \begin{pmatrix} x_k \\ y_k \end{pmatrix} + \alpha_k \begin{pmatrix} p_k \\ q_k \end{pmatrix} \qquad\qquad \mathbb{x}_{k+1} = \mathbb{x}_k + \alpha_k \mathbb{p}_k$$

$$\mathbb{r}_{k+1} = \mathbb{r}_k - \alpha_k \mathbb{A} \begin{pmatrix} p_k \\ q_k \end{pmatrix} = \begin{pmatrix} 0 \\ r_{k+1} \end{pmatrix} \qquad\qquad \mathbb{r}_{k+1} = \mathbb{r}_k - \alpha_k \mathbb{A}\mathbb{p}_k$$

$$\beta_k = \frac{\left(\begin{pmatrix} 0 \\ r_{k+1} \end{pmatrix}, \mathbb{P}^{-1}\begin{pmatrix} 0 \\ r_{k+1} \end{pmatrix} \right)}{\left(\begin{pmatrix} 0 \\ r_k \end{pmatrix}, \mathbb{P}^{-1}\begin{pmatrix} 0 \\ r_k \end{pmatrix} \right)} \qquad\qquad \beta_k = \frac{(\mathbb{r}_{k+1}, \mathbb{P}^{-1}\mathbb{r}_{k+1})}{(\mathbb{r}_k, \mathbb{P}^{-1}\mathbb{r}_k)}$$

$$\begin{pmatrix} p_{k+1} \\ q_{k+1} \end{pmatrix} = \mathbb{P}^{-1}\begin{pmatrix} 0 \\ r_{k+1} \end{pmatrix} + \beta_k \begin{pmatrix} p_k \\ q_k \end{pmatrix} \qquad\qquad \mathbb{p}_{k+1} = \mathbb{P}^{-1}\mathbb{r}_{k+1} + \beta_k \mathbb{p}_k$$

end

---

the solution x_* with $\|x_* - x_{k+1}\|_A \to 0$. However, when using Algorithm 8.4, in general, the vector y_{k+1} does not converge to y_* which is also reflected in the non-convergence of the residual vector $\mathbb{r}_{k+1} = \begin{pmatrix} r_{k+1} \\ 0 \end{pmatrix}$. Nevertheless, under appropriate scaling we can achieve also the convergence of y_{k+1} to y_*. Since the errors $x_* - x_{k+1}$ can be represented using some polynomials p_{k+1} of degree $k + 1$ as $x_* - x_{k+1} = p_{k+1}(ZZ^T AZZ^T)(x_* - x_0)$, the vectors r_{k+1} can be expressed in the form (7.36) using the same polynomial but in terms of the nonsymmetric matrix $AZZ^T + B(B^T B)^{-1}B^T$. The vectors r_{k+1} will converge if the spectrum of $AZZ^T + B(B^T B)^{-1}B^T$ coincides with the nonzero eigenvalues of $ZZ^T AZZ^T$. Fortunately, the nonzero eigenvalues of these two matrices differ only in the unit eigenvalues. If we find a scaling constant $1/\tau > 0$ such that $\{1\} \in \text{sp}(ZZ^T 1/\tau AZZ^T)$, then necessarily the residual vector r_{k+1} will converge with $\|\mathbb{r}_{k+1}\| = \left\| \begin{pmatrix} r_{k+1} \\ 0 \end{pmatrix} \right\| \to 0$.

There are several possible ways to avoid the misconvergence of approximate solutions y_{k+1} that may work in practical situations:

- Scaling by a constant $1/\tau > 0$ so that the unit eigenvalue belongs to the convex hull of nonzero eigenvalues of the matrix $ZZ^T (1/\tau)AZZ^T$ and, so it belongs

---

Algorithm 8.4 Preconditioned CG with the constraint preconditioner (7.31) in the null-space method

---

$$\text{choose} \begin{pmatrix} x_0 \\ y_0 \end{pmatrix}, \quad \mathbb{r}_0 = \mathbb{b} - \mathbb{A} \begin{pmatrix} x_0 \\ y_0 \end{pmatrix} = \begin{pmatrix} r_0 \\ 0 \end{pmatrix}$$

$$\begin{pmatrix} p_0 \\ q_0 \end{pmatrix} = \mathbb{P}^{-1} \mathbb{r}_0 = \mathbb{P}^{-1} \begin{pmatrix} r_0 \\ 0 \end{pmatrix}$$

for $k = 0, 1, \ldots$

$$\alpha_k = \frac{\left(\begin{pmatrix} r_k \\ 0 \end{pmatrix}, \mathbb{P}^{-1} \begin{pmatrix} r_k \\ 0 \end{pmatrix} \right)}{\left(\mathbb{A} \begin{pmatrix} p_k \\ q_k \end{pmatrix}, \begin{pmatrix} p_k \\ q_k \end{pmatrix} \right)} \qquad\qquad \alpha_k = \frac{(r_k, \mathbb{P}^{-1} r_k)}{(\mathbb{A} p_k, p_k)}$$

$$\begin{pmatrix} x_{k+1} \\ y_{k+1} \end{pmatrix} = \begin{pmatrix} x_k \\ y_k \end{pmatrix} + \alpha_k \begin{pmatrix} p_k \\ q_k \end{pmatrix} \qquad\qquad x_{k+1} = x_k + \alpha_k p_k$$

$$\mathbb{r}_{k+1} = \mathbb{r}_k - \alpha_k \mathbb{A} \begin{pmatrix} p_k \\ q_k \end{pmatrix} = \begin{pmatrix} r_{k+1} \\ 0 \end{pmatrix} \qquad\qquad \mathbb{r}_{k+1} = \mathbb{r}_k - \alpha_k \mathbb{A} p_k$$

$$\beta_k = \frac{\left(\begin{pmatrix} r_{k+1} \\ 0 \end{pmatrix}, \mathbb{P}^{-1} \begin{pmatrix} r_{k+1} \\ 0 \end{pmatrix} \right)}{\left(\begin{pmatrix} r_k \\ 0 \end{pmatrix}, \mathbb{P}^{-1} \begin{pmatrix} r_k \\ 0 \end{pmatrix} \right)} \qquad\qquad \beta_k = \frac{(r_{k+1}, \mathbb{P}^{-1} r_{k+1})}{(r_k, \mathbb{P}^{-1} r_k)}$$

$$\begin{pmatrix} p_{k+1} \\ q_{k+1} \end{pmatrix} = \mathbb{P}^{-1} \begin{pmatrix} r_{k+1} \\ 0 \end{pmatrix} + \beta_k \begin{pmatrix} p_k \\ q_k \end{pmatrix} \qquad\qquad \mathbb{p}_{k+1} = \mathbb{P}^{-1} r_{k+1} + \beta_k \mathbb{p}_k$$

end

---

also to the convex hull of eigenvalues of the matrix $Z^T (1/\tau) A Z$ and considering the equivalent saddle-point system $\mathbb{A}(\tau) \mathbb{x}(\tau) = \mathbb{b}(\tau)$ or

$$\begin{pmatrix} \frac{1}{\tau} A & B \\ B^T & 0 \end{pmatrix} \begin{pmatrix} x \\ \frac{1}{\tau} y \end{pmatrix} = \begin{pmatrix} \frac{1}{\tau} c \\ 0 \end{pmatrix}. \tag{8.21}$$

This can be achieved by taking any vector $v \in \mathcal{R}^m$ such that $\| Z^T v \| \neq 0$ and by computing the scaling factor as $\tau = (ZZ^T v, A ZZ^T v)$.

- If A is symmetric positive definite, then the eigenvalues of the matrix $(\text{diag}(A))^{-1/2} A (\text{diag}(A))^{-1/2}$ are either all equal to 1 or there exists at least one eigenvalue less than 1 and one eigenvalue which is greater than 1 so that they are contained in a nontrivial positive interval strictly including the number 1.
- Since $\mathbb{A}$ and $\mathbb{P}$ are only symmetric indefinite, the conjugate gradient direction q_k may not exist in Algorithm 8.4, and a different direction vector q_k must be chosen. One possible choice is to compute q_k so that $\| \mathbb{r}_{k+1} \| = \| r_{k+1} \|$ is minimized via solving the least squares problem

$$y_{k+1} = y_k + \arg\min_{y \in \mathcal{R}^n} \| r_k - B y \| = y_k + (B^T B)^{-1} B^T r_k.$$

1/τ	sp($\mathbb{A}(\tau)\mathbb{P}^{-1}$)
1/100	[0.0207, 0.0586] ∪ {1}
1/10	[0.2067, 0.5856] ∪ {1}
1/4	**[0.5170, 1.4641]**
1	{1} ∪ [2.0678, 5.8563]
4	{1} ∪ [8.2712, 23.4252]

Table 8.4 Inclusion sets for eigenvalues of the matrix $\mathbb{A}(\tau)\mathbb{P}^{-1}$ scaled as in (8.21) for several values of the parameter τ

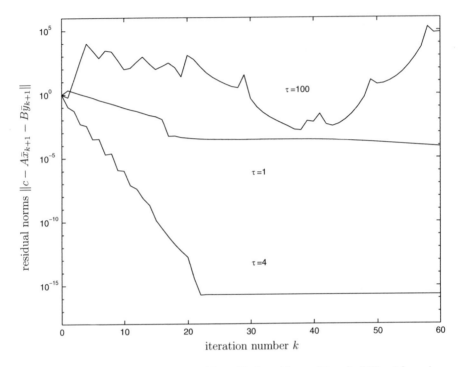

Fig. 8.14 Preconditioned CG: the norms of the residual $c - A\bar{x}_{k+1} - B\bar{y}_{k+1}$ (solid lines) for various values of the scaling parameter τ

In the following we again consider the model problem (8.1), where we set $m = 25$ and $n = 5$. The eigenvalues of the matrix A and so the eigenvalues of the matrix $Z^T A Z$ belong to the interval [2.0146, 5.9854]. In Table 8.4 we give the inclusion sets for the eigenvalues of the matrix $\mathbb{A}(\tau)\mathbb{P}^{-1}$ scaled as in (8.21) for several values of the parameter τ. Clearly, $\tau = 1$ gives the original preconditioned matrix $\mathbb{A}\mathbb{P}^{-1}$, while $\tau > 1$ moves the inclusion set toward zero. The value $\tau = 4$ is optimal in the sense that it corresponds to the scaling by diagonal $(\mathrm{diag}(A))^{-1/2}A(\mathrm{diag}(A))^{-1/2}$, and thus the inclusion set contains 1.

In Fig. 8.14 we show the norm of the residual $c - A\bar{x}_{k+1} - B\bar{y}_{k+1}$ for $\tau = 1$, $\tau = 4$, and $\tau = 10$. Figure 8.15 shows the A-norm of the error $x_* - \bar{x}_{k+1}$ for the same values of τ. Indeed, for $\tau = 4$ both the residual norm and the error norm

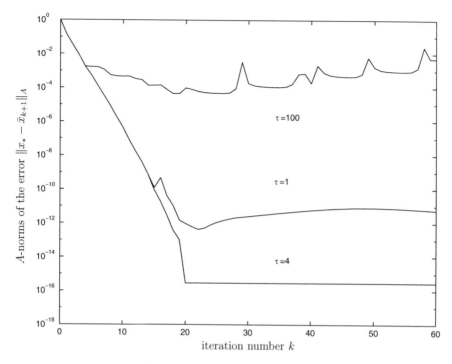

Fig. 8.15 Preconditioned CG: the energy norms of the errors $x_* - \bar{x}_{k+1}$ (solid lines) for various values of the scaling parameter τ

converge to the machine precision level, while for the unscaled case $\tau = 1$ the residual norm does not decrease at the same rate as the A-norm of the error and for $\tau = 100$ the residual norm seems to diverge. According to theory, the A-norm of the error $x_* - x_{k+1}$ should converge for all values of τ, but Fig. 8.15 shows that unappropriate scaling may influence the maximum attainable accuracy of computed approximate solutions. For $\tau = 100$ we see a complete failure of the method due to rounding errors. The maximum attainable accuracy of the preconditioned CG method, measured in terms of the A-norm of the error $x_* - \bar{x}_{k+1}$, was analyzed in [71]. It was shown that a proper scaling not only ensures the convergence of the residual norm in exact arithmetic, but it also leads to satisfactory level of maximum attainable accuracy. For details we refer to Section 10 in [71].

Chapter 9
Modeling of Polluted Groundwater Flow in Porous Media

This chapter is devoted to the case study that comes from a real-world application of groundwater flow modeling in the area of Stráž pod Ralskem in northern Bohemia. We give some basic facts about the uranium mining in northern Bohemia together with a short description of main activities of the Department of Mathematical Modeling at the state enterprise DIAMO in Stráž pod Ralskem.

As an example of models used in Stráž pod Ralskem, we consider the potential fluid flow in porous media described via Darcy's law and the continuity equation discretized by the mixed-hybrid finite element method using trilateral prismatic elements that reflect well the complex geological structure with layers of stratified rocks and sedimented minerals.

We consider several approaches for solving the resulting large-scale saddle-point problems. We discuss the iterative solution of the original saddle-point problem using the MINRES method, the approach based on the successive block reductions with subsequent iterative solution of the Schur complement systems and the null-space approach based on iterative solution of systems projected onto certain null-spaces. We look at the convergence rate of iterative methods used in these schemes and give bounds for their asymptotic convergence factor in terms of the discretization parameter h. Theoretical results on the convergence rate and some numerical tests were published in a series of papers [4, 57, 58, 60]. Several implementations and codes written by the authors were used in GWS software package developed in DIAMO, s. e. It follows that the best asymptotic convergence factor of all main approaches is the same. Therefore, when attempting to compare their computational efficiency, we must take into account not only the number of iterations but also the initial overhead of the reduction to Schur complement systems, the computation of the null-space basis, and the back-substitution as well as other transformations such as prescaling or preconditioning. Extensive and systematic numerical experiments are out of the scope of this textbook. We consider

© Springer Nature Switzerland AG 2018
M. Rozložník, *Saddle-Point Problems and Their Iterative Solution*,
Nečas Center Series, https://doi.org/10.1007/978-3-030-01431-5_9

only the unpreconditioned iterative methods here, but in practice they should be applied together with efficient preconditioning that ideally leads to convergence rate independent of the discretization. Examples of optimal preconditioning for Raviart-Thomas mixed formulation of second-order elliptic problem (1.8) can be found in [68]. Another efficient approach for this type of problems is to use the algebraic multigrid algorithm for the solution of linear systems arising from discontinuous Galerkin discretizations [9] or, in this application, for the solution of reduced Schur complement of null-space projected systems.

9.1   Uranium Mining in Northern Bohemia

Geological exploration of uranium in Czechoslovakia started in 1946, and it led to mining activities at 66 uranium deposits. The geographical location of the uranium deposit Stráž pod Ralskem in the Czech Republic is depicted in Fig. 9.1. Two basic mining technologies were used there: the classical underground mining and in situ chemical leaching. The classical deep mines were operated in the years 1966–1993 and the in situ acidic leaching in the years 1968–1996, and since 1996 the deposit is in the phase of remediation under the supervision of DIAMO, state enterprise in Stráž pod Ralskem. During this period, 11,600 tons of uranium were produced by

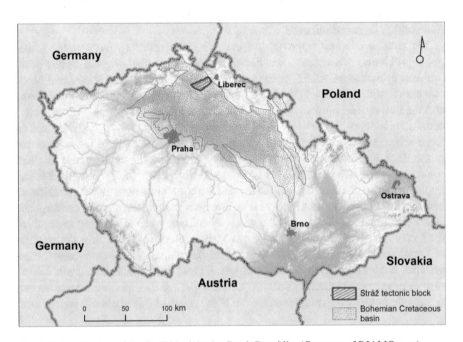

Fig. 9.1 Localization of the Stráž block in the Czech Republic. (Courtesy of DIAMO, s. e.)

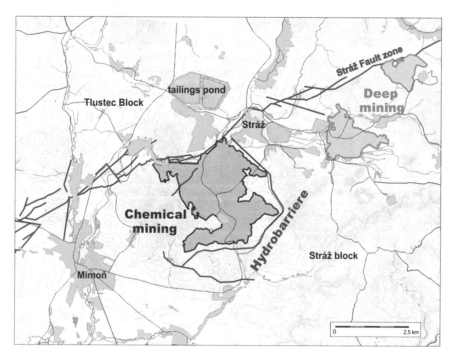

Fig. 9.2 The uranium mining area in the North Bohemia. (Courtesy of DIAMO, s. e.)

deep mining, and 15,800 tons of uranium were produced by in situ leaching. The basic situation map of the mining area can be found in Fig. 9.2. The near existence of two totally different mining methods led to unique hydraulic conditions. The hydraulic barrier was installed with the pressurized injection of clean water keeping the level of groundwater in leaching fields close to the surface and protecting the deep mines from the floods with a contaminated solution. The top view with the location of leaching fields and hydrological barrier is given in Fig. 9.3. During the mining activities, 15,000 wells were drilled, and drainage channels were built for injection of sulfur acid and pumping out the uranium ore solution. The scheme of remote-controlled system of injection and recovery wells is depicted in Fig. 9.4. The total area of leaching fields is 628 ha; $4 \cdot 10^6$ tons of H_2SO_4 and $3 \cdot 10^5$ tons of HNO_3 were injected in sandstone area causing the ecological load of $374 \cdot 10^6$ m^3 of contaminated water in Cenomanian aquifer within the area of 27.3 km^2. The cross-section of the area with contamination can be seen in Fig. 9.5. Besides the liquidation of shafts, decommission of boreholes and surface installations, the main objective of remedial activities is to restore the rock environment to a condition that will guarantee the incorporation of leaching fields into a public ecosystem allowing their sustainable development in the framework of valid urban plans. These activities include pumping out the contaminants from the underground and their processing to industrially usable and ecologically storable products as well as in

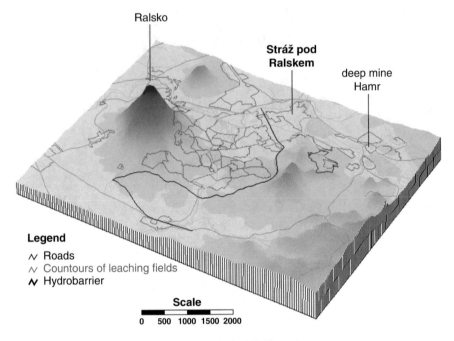

Fig. 9.3   Location of leaching fields. (Courtesy of DIAMO, s. e.)

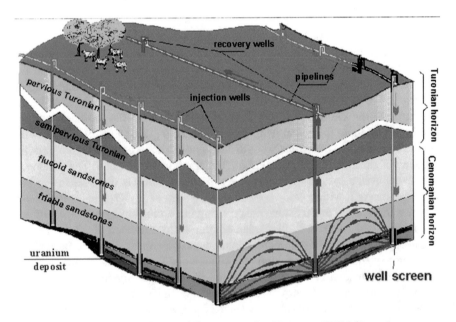

Fig. 9.4   Remote-controlled injection and recovery wells. (Courtesy of DIAMO, s. e.)

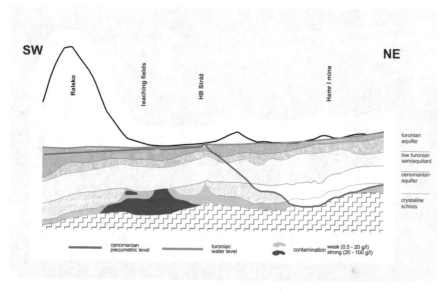

Fig. 9.5 Schematic cross-section of the area around Stráž pod Ralskem. (Courtesy of DIAMO, s. e.)

situ immobilization of contaminants fixing the hydrogeological conditions in the whole area.

Mathematical modeling is one of the essential tools for remediation management. The expertise of the Department of Mathematical Modelling in DIAMO, s. e., lies in applications such as groundwater flow, transport of contaminants, design of remediation scenarios, flooding scenarios of deep uranium mines, chemical leakage from waste ponds, fractured rock flow, and evaluation of radioactive waste deposits. Various mathematical models were developed and used including structural and description models that provide input data for other models, computational models of flow and transport, thermodynamical and kinetical models of chemical processes and reactions, and optimization models making support and tools for economical decisions. For illustration, in Figs. 9.6, 9.7 and 9.8 we show the output from the GWS software developed in DIAMO with the computed scenario of the underground contamination within three consecutive remediation time periods after termination of in situ leaching. For details we refer to the technical report [50].

9.2   Potential Fluid Flow Problem in Porous Media

The potential fluid flow problem in porous media is one of the most important mathematical problems that arise in computational models developed at DIAMO. The fluid flow in porous medium is usually described by the Darcy's law (1.12)

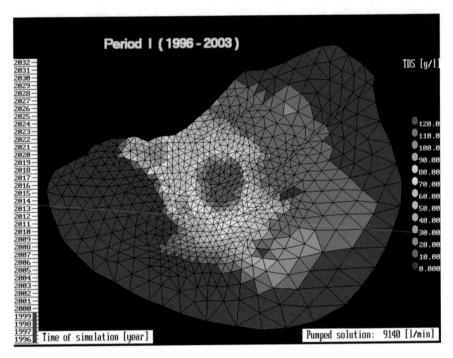

Fig. 9.6 Computed contamination of the leaching area in the period 1996–2003. (Courtesy of DIAMO, s. e.)

which gives a relation between the potential head (fluid pressure) p and the fluid velocity $\mathbf{u}$ which also satisfies the continuity equation representing the mass conservation law in the domain together with some prescribed boundary conditions. The mixed finite element discretization is a very efficient and popular approximation technique for this type of problems, especially when an accurate approximation of the fluid velocity is required. However, in some cases the system matrix after the mixed finite element discretization may become ill-conditioned, and the hybridization seems to be one of the possible strategies to avoid this problem. In addition, the local conservation property of the mixed-hybrid discretization models well the transport phenomena, and as we will see later, the saddle-point matrices in linear algebraic systems resulting from hybridization have a rather transparent and simple sparsity structure.

The mixed-hybrid weak formulation is given and discussed in [57] (see also [59]). We denote the collection of subdomains of the domain Ω by $\mathcal{E}_h$, where h is the discretization parameter given as $h = \max_{e \in \mathcal{E}_h} \{\text{diam } e\}$, and the collection of faces of subdomains $e \in \mathcal{E}_h$ that are not adjacent to the boundary $\partial \Omega_D$ used in hybridization is denoted by $\Gamma_h = \cup_{e \in \mathcal{E}_h} \partial e - \partial \Omega_D$. Whenever it applies, the restriction of any function φ defined on Ω on the subdomain $e \in \mathcal{E}_h$ is denoted by $\varphi|_e$. Indeed, the weak formulation of the problem (1.12) on the discretization $\mathcal{E}_h$ of

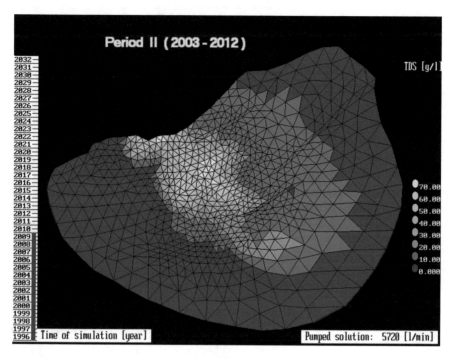

Fig. 9.7 Computed contamination of the leaching area in the period 2003–2012. (Courtesy of DIAMO, s. e.)

the domain Ω can be stated as follows:

Find $(\mathbf{u}_*, p_*, \lambda_*) \in \mathbf{H}(div, \mathcal{E}_h) \times L^2(\Omega) \times H_D^{\frac{1}{2}}(\Gamma_h)$ such that

$$\sum_{e \in \mathcal{E}_h} \left\{ \left(A^{-1}\mathbf{u}_*|_e, \mathbf{u}|_e \right)_e - (p_*|_e, \nabla \cdot \mathbf{u}|_e)_e \right. \tag{9.1}$$

$$\left. + (\lambda_*|_e, \mathbf{n}|_e \cdot \mathbf{u}|_e)_{\partial e \cap \Gamma_h} \right\} = \sum_{e \in \mathcal{E}_h} (p_D|_e, \mathbf{n}|_e \cdot \mathbf{u}|_e)_{\partial e \cap \partial \Omega_D},$$

$$\sum_{e \in \mathcal{E}_h} (\nabla \cdot \mathbf{u}_*|_e, p|_e)_e = \sum_{e \in \mathcal{E}_h} (f|_e, p|_e)_e,$$

$$\sum_{e \in \mathcal{E}_h} (\mathbf{n}|_e \cdot \mathbf{u}_*|_e, \lambda|_e)_{\partial e} = \sum_{e \in \mathcal{E}_h} (u_N|_e, \lambda|_e)_{\partial e \cap \partial \Omega_N}$$

for all $(\mathbf{u}, p, \lambda) \in \mathbf{H}(div, \mathcal{E}_h) \times L^2(\Omega) \times H_D^{\frac{1}{2}}(\Gamma_h)$, where $\mathbf{H}(div, \mathcal{E}_h)$ is the space of square integrable vector functions $\mathbf{u} \in \mathbf{L}^2(\Omega)$ with square integrable divergences

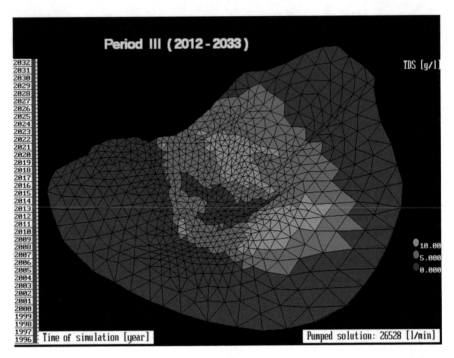

Fig. 9.8 Computed contamination of the leaching area in the period 2012–2033. (Courtesy of DIAMO, s. e.)

on every subdomain $e \in \mathcal{E}_h$, i.e.,

$$\mathbf{H}(div, \mathcal{E}_h) = \left\{ \mathbf{u} \in \mathbf{L}^2(\Omega) \mid \nabla \cdot \mathbf{u}|_e \in L^2(e), \; \forall e \in \mathcal{E}_h \right\},$$

and where $H_D^{\frac{1}{2}}(\Gamma_h)$ is the space of traces

$$H_D^{\frac{1}{2}}(\Gamma_h) = \left\{ \lambda : \Gamma_h \rightarrow \mathcal{R} \mid \exists \varphi \in H_D^1(\Omega), \; \lambda = \varphi|_{\Gamma_h} \right\}.$$

The space $H_D^1(\Omega)$ is defined as

$$H_D^1(\Omega) = \left\{ \varphi : \Omega \rightarrow \mathcal{R} \mid \varphi \in L^2(\Omega), \; \nabla\varphi \in L^2(\Omega), \; \varphi|_{\partial\Omega} = 0 \text{ on } \partial\Omega_D \right\},$$

where $\varphi|_{\partial\Omega}$ and $\varphi|_{\Gamma_h}$ is the trace of the function φ on the boundary $\partial\Omega$ and on the structure of faces Γ_h, respectively. For details we refer to Section 2 in [57]; see also [47].

Among the specifics making the application in northern Bohemia unique belong the impermeable (generally nonparallel) bottom and top layers and the fact that the modeled area covers approximately $120\,\text{km}^2$, while its vertical thickness is up to $200\,\text{m}$ (see the map of the area in Fig. 9.2 and its cross-section in Fig. 9.5). The layers of stratified rocks with substantially different physical properties must be modeled by an appropriate discretization that reflects complex geological structure of sedimented minerals. Therefore, trilateral prismatic elements with vertical faces and generally nonparallel bases are used together with the low-order Raviart-Thomas approximation of velocity, fluid pressure, and Lagrange multipliers from hybridization. The velocity function $\mathbf{u}$ will be then approximated by vector functions $\mathbf{u}_h$ linear on every element $e \in \mathcal{E}_h$ from the Raviart-Thomas space

$$\mathbf{RT}^0_{-1}(\mathcal{E}_h) = \{\mathbf{u}_h \in \mathbf{L}^2(\Omega);\ \mathbf{u}_h|_e \in \mathbf{RT}^0(e),\ \forall e \in \mathcal{E}_h\}, \tag{9.2}$$

where $\mathbf{RT}^0(e)$ denotes the five-dimensional space spanned by linear functions that are orthormal with respect to set of functionals determined by the geometry of prismatic elements. Further we denote the space of constant functions on each element $e \in \mathcal{E}_h$ by $M^0(e)$ and denote the space of constant functions on each face $f \in \Gamma_h$ by $M^0(f)$. The piezometric potential p is approximated by the space which consists of element-wise constant functions

$$M^0_{-1}(\mathcal{E}_h) = \{\varphi_h \in L^2(\Omega);\ \varphi_h|_e \in M^0(e),\ \forall e \in \mathcal{E}_h\}. \tag{9.3}$$

The Lagrange multipliers λ are approximated by the space of all functions λ_h constant on every face f from Γ_h

$$M^0_{-1}(\Gamma_h) = \{\lambda_h \in L^2(\Omega);\ \Gamma_h \to R;\ \lambda_h|_f \in M^0(f),\ \forall f \in \Gamma_h\}. \tag{9.4}$$

For details we refer to [57] and [58], where the appropriate basis functions are defined and the existence and uniqueness of the approximate solution to the discretized problem are proved.

If we denote by $ne = |\mathcal{E}_h|$ the number of elements used in the discretization, by nif the number of interior inter-element faces, and by nnc the number of faces with prescribed Neumann conditions ($nif + nnc = |\Gamma_h|$), then from the abovementioned Raviart-Thomas discretization, we obtain the system of linear algebraic equations in the form

$$\begin{pmatrix} A & B_1 & B_2 & B_3 \\ B_1^T & 0 & 0 & 0 \\ B_2^T & 0 & 0 & 0 \\ B_3^T & 0 & 0 & 0 \end{pmatrix} \begin{pmatrix} x \\ y_1 \\ y_2 \\ y_3 \end{pmatrix} = \begin{pmatrix} c \\ d_1 \\ d_2 \\ d_3 \end{pmatrix}, \tag{9.5}$$

where $x \in \mathcal{R}^{5 \cdot ne}$, $y_1 \in \mathcal{R}^{ne}$, $y_2 \in \mathcal{R}^{nif}$, and $y_3 \in \mathcal{R}^{nnc}$. The matrix block $A \in \mathcal{R}^{5 \cdot ne, 5 \cdot ne}$ represents the discrete form of the hydraulic permeability tensor, the matrix block $B_1^T \in \mathcal{R}^{ne,5 \cdot ne}$ enforces the continuity equation on every element, the matrix block $B_2^T \in \mathcal{R}^{nif,5 \cdot ne}$ represents the continuity equation across all interior inter-element faces, and the matrix block $B_3^T \in \mathcal{R}^{nif,5 \cdot ne}$ stands for the fulfillment of all Neumann conditions prescribed on the boundary of the domain. It is clear from (9.5) that substituting for $B = (B_1 \ B_2 \ B_3)$, $y = \left(y_1^T \ y_2^T \ y_3^T\right)^T$, and $d = \left(d_1^T \ d_2^T \ d_3^T\right)^T$ we have the standard saddle-point problem (1.1) with (1.3), where $m = 5 \cdot ne$ and $n = ne + nif + nnc$. In addition, in most cases we have that $d = 0$ and thus the right-hand side has the form (1.4).

From the algebraic point of view, it was shown in [58] that as a result of the hybridization of the scheme, the matrix block A is symmetric positive definite and it is block diagonal with 5-by-5 diagonal blocks. In addition, the matrix block B_1 is essentially the face-element incidence matrix with values equal to -1. Since each element has exactly 5 faces, we have that $B_1^T B_1 = 5I$, and thus all singular values of B_1 are equal to $\sqrt{5}$. Hence, up to the column scaling, the matrix B_1 is orthogonal. The columns of the matrix B_2 contain only two nonzero entries equal to 1 and satisfy the condition $B_2^T B_2 = 2I$. Therefore, all singular values of B_2 are equal to $\sqrt{2}$. The matrix block B_3 is the face-condition incidence matrix of the element faces with Neumann boundary conditions. So, it is an orthogonal matrix satisfying $B_3^T B_3 = I$. The nonzero entries of the matrix $\mathbb{A}$ from a small model problem are depicted in Fig. 9.9.

In the following we describe a model potential fluid flow problem that is used in all numerical examples in subsequent three subsections. We consider the potential fluid flow problem with a constant tensor of hydraulic permeability in a cubic domain with Dirichlet conditions prescribed on the top and on the bottom of the domain together with Neumann boundary conditions imposed on the rest of the boundary. The choice of boundary conditions is motivated by our application of a domain between two impermeable layers. We report the results from a uniformly regular mesh refinement with prismatic elements for different values of the discretization parameter h. For the cubic domain, we have then $ne = 2/h^3$, and the dimension of the resulting saddle-point matrix $\mathbb{A}$ is approximately $m + n \sim 22/h^3$. The inclusion sets for the extremal eigenvalues or singular values of matrices A, B, $\mathbb{A}$, $-\mathbb{A} \backslash A$ or $Z^T A Z$ were calculated using the double precision LAPACK and NAG subroutines and using the symmetric Lanczos algorithm and subsequent use of LAPACK solvers on the resulting tridiagonal matrix. For details of numerical examples, we refer to papers [4, 58] and [59].

Note that the time-dependent real-world applications in DIAMO, s. e., use three-dimensional unstructured meshes of the order from $2 \cdot 10^5$ to $8 \cdot 10^5$ elements that do not allow a uniform mesh refinement, and they lead to saddle-point problems (9.5) of order $10^6 - 10^7$ that are moreover solved at each time step of the chosen time discretization.

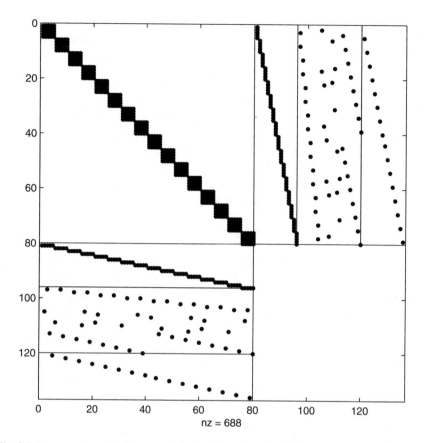

Fig. 9.9 Nonzero elements of the matrix obtained from the mixed-hybrid discretization of a model problem

9.3 Minimal Residual Method Applied to Saddle-Point Problem

By analysis of the entries in the block diagonal and symmetric positive definite matrix A, it was shown in Subsection 2.1 of [58] that its eigenvalues are included in the interval

$$\mathrm{sp}(A) \subset \left[\frac{\mu_1}{h}, \frac{\mu_2}{h}\right], \tag{9.6}$$

where h is a discretization parameter (mesh size) and where μ_1 and μ_2 are positive constants independent of the discretization. These constants, however, depend on the properties of the original continuous problem such as hydraulic permeability tensor and the geometry of a given domain. The condition number of A does not depend on h, and it is equal to $\kappa(A) = \mu_2/\mu_1$.

Although the matrix blocks B_1, B_2, and B_3 are up to the column scaling orthogonal, the whole off-diagonal block $B = (B_1\ B_2\ B_3)$ is no longer orthogonal, and it can be fairly ill-conditioned. The standard approach used in the conformal mixed finite element method to estimate the minimal singular value of the off-diagonal block B is to apply the discrete version of the inf-sup (Babuška-Brezzi) condition (2.17). The situation in our application with the finite-dimensional subspaces (9.2), (9.3) and (9.4) used for discretization of (9.1) is more complicated, the inf-sup condition cannot be used, and the spectral properties of B must be analyzed directly from its structure of nonzero entries. The singular values of B were therefore analyzed using graph theoretical tools in Subsection 2.2 of [58]. Assuming at least one Dirichlet condition imposed on the boundary of a domain, it was shown using the tools from graph theory that they are included in the interval

$$\mathrm{sv}(B) \subset [\mu_3 h, \mu_4], \tag{9.7}$$

where μ_3 and $\mu_4 = \sqrt{5} + \sqrt{2}$ are positive constants independent of the system parameters but dependent on physical and geometrical properties of a given domain. The inclusion sets for the computed extremal eigenvalues of A and the extremal singular values of B are shown in Table 9.1, and they are in good agreement with (9.6) and (9.7).

According to the result of Rusten and Winther formulated in Proposition 3.5, the eigenvalues of the saddle-point matrix $\mathbb{A}$ can be related to the eigenvalues of A and to the singular values of B. Thus using (3.16) together with (9.6) and (9.7), we get the inclusion set for the eigenvalues of the matrix $\mathbb{A}$ in the form

$$\mathcal{G}(h) = \left[-\frac{\mu_4^2}{\mu_1}h, -\frac{\mu_3^2}{\mu_2}h^3 \right] \cup \left[\frac{\mu_1}{h}, \frac{\mu_2}{h} \right]. \tag{9.8}$$

The condition number of $\mathbb{A}$ is approximately of the order of $\frac{\mu_2^2}{\mu_3^2 h^4}$. Note that since $h \to 0$, in (9.8) we have omitted higher-order terms, and we will use the same approach throughout this subsection. For practically feasible values of discretization parameter h in our application, the realistic conditions for the constants μ_1, μ_2, μ_3, and μ_4 are represented by $\mu_1/h \ll \mu_4$, $\mu_2/h \ll \mu_4$, and $\mu_2/h \gg \mu_3 h$ (see the paper [58]). Then, assuming that the above-stated conditions are still valid, for the

Table 9.1 Inclusion sets for the spectrum of A and for singular values of B

Mesh size h	Spectrum of A	Singular values of B
1/2	[0.11e-2, 0.66e-2]	[0.276, 2.577]
1/3	[0.16e-2, 0.10e-1]	[0.197, 2.611]
1/4	[0.22e-2, 0.13e-1]	[0.152, 2.624]
1/6	[0.33e-2, 0.19e-1]	[0.104, 2.635]
1/8	[0.44e-2, 0.26e-1]	[0.079, 2.639]

Table 9.2 Inclusion sets for the spectrum of the saddle-point matrix $\mathbb{A}$

Mesh size h	Negative part	Positive part
1/2	$[-2.57, -0.27]$	$[0.13\text{e-}2, 2.57]$
1/3	$[-2.60, -0.19]$	$[0.18\text{e-}2, 2.61]$
1/4	$[-2.62, -0.15]$	$[0.23\text{e-}2, 2.62]$
1/6	$[-2.63, -0.13]$	$[0.34\text{e-}2, 2.63]$
1/8	$[-2.63, -0.10]$	$[0.44\text{e-}2, 2.64]$

eigenvalues of $\mathbb{A}$, we have the inclusion set

$$\mathcal{G}(h) = \left[-\mu_4, -\frac{\mu_3^2}{\mu_2}h^3 \right] \cup \left[\frac{\mu_1}{h}, \mu_4 \right] \tag{9.9}$$

leading to the condition number $\kappa(\mathbb{A})$ of the order $\frac{\mu_2\mu_4}{\mu_3^2 h^3}$. In Table 9.2 we give the inclusion sets for the spectrum of $\mathbb{A}$ computed for several values of the discretization parameter h. The results are in a good agreement with (9.9).

If we consider the diagonal scaling of the saddle-point matrix (1.3) in the system (1.1), then we get the scaled matrix

$$\begin{pmatrix} h^{1/2}I & 0 \\ 0 & h^{-1/2}I \end{pmatrix} \begin{pmatrix} A & B \\ B^T & 0 \end{pmatrix} \begin{pmatrix} h^{1/2}I & 0 \\ 0 & h^{-1/2}I \end{pmatrix} = \begin{pmatrix} hA & B \\ B & 0 \end{pmatrix} \tag{9.10}$$

with the diagonal block equal to hA and with the off-diagonal block B which remains untouched. The spectrum of hA lies in the interval $[c_1, c_2]$ that is independent of h. The spectrum of the scaled matrix (9.10) is included in

$$\mathcal{G}(h) = \left[\frac{1}{2}(\mu_1 - \sqrt{\mu_1^2 + 4\mu_4^2}), -\mu_3^2\mu_2^{-1}h^2 \right] \cup \left[\mu_1, \frac{1}{2}(\mu_2 + \sqrt{\mu_2^2 + 4\mu_4^2}) \right]. \tag{9.11}$$

It is clear from (9.11) that the conditioning of the scaled matrix (9.10) is approximately of the order $\frac{\mu_2}{2\mu_3^2 h^2}(\mu_2 + \sqrt{\mu_2^2 + 4\mu_4^2})$.

The most straightforward approach to solve saddle-point problems is to apply the MINRES method to the system (1.1) without any preprocessing or a problem reduction. As noted in (6.34), the convergence of the MINRES method depends on the eigenvalue distribution, and the relative residual norm of MINRES can be estimated via the best minimal polynomial approximation on the spectrum of the matrix $\mathbb{A}$. This discrete approximation problem is then relaxed to the polynomial approximation on a continuous inclusion set given by two disjoint intervals $[-\alpha, -\beta] \cup [\gamma, \delta]$, where $0 < \beta < \alpha$ and $0 < \gamma < \delta$. The optimal polynomial is known only in special cases such as (6.35). In practical situations such as (9.8), (9.9), and (9.11), one can estimate only the asymptotic convergence factor defined by (6.36). Indeed, the rightmost term in (6.36) is estimated further by

$\lim_{k \to \infty} \left(\min_{\substack{p \in P_k \\ p(0)=1}} \max_{\lambda \in \mathcal{G}(h)} |p(\lambda)| \right)^{\frac{1}{k}}$, where $\mathcal{G}(h)$ denotes an inclusion set for
the spectrum of $\mathbb{A}$. For details of such analysis, we refer to [82]. Using the approach
of Wathen, Fischer, and Silvester, it was shown in Subsection 3.2 of [58] that the
asymptotic convergence factor with the inclusion set (9.8) is bounded as

$$\lim_{k \to \infty} \left(\frac{\|r_k\|}{\|r_0\|} \right)^{\frac{1}{k}} \leq 1 - \xi_1 h^2, \qquad (9.12)$$

where ξ_1 is a positive constant dependent on the constants μ_1, μ_2, μ_3, and μ_4 and
independent of the parameter h. However, considering the more realistic inclusion
set (9.9) or the inclusion set (9.11) after the scaling (9.10), we get the upper bound

$$\lim_{k \to \infty} \left(\frac{\|r_k\|}{\|r_0\|} \right)^{\frac{1}{k}} \leq 1 - \xi_2 h. \qquad (9.13)$$

In both cases, the constant ξ_2 is positive and depends only on the constants μ_1,
μ_2, μ_3, and μ_4. The bound (9.12) indicates that the asymptotic rate of convergence
of MINRES applied to (1.1) can be slow for very small values of h, but it can
be improved significantly by a diagonal scaling (9.10). Then, the bound for the
asymptotic convergence factor depends at most linearly on h. This is also true for
realistic problems in our application with the inclusion set (9.9). In Table 9.3 we
give the number of iteration steps of MINRES and the average contraction factors
computed as $\left(\frac{\|r_k\|}{\|r_0\|} \right)^{1/k}$ after relative residual norm reduction of 10^{-4}, 10^{-8}, and
10^{-12} for several values of the discretization parameter h. The convergence behavior
of the MINRES method on these systems is shown in Fig. 9.10.

Table 9.3 Number of iteration steps and average contraction factor of MINRES after the relative residual norm reduction of 10^{-4}, 10^{-8}, and 10^{-12}	h	$\frac{\|r_k\|}{\|r_0\|} = 10^{-4}$	$\frac{\|r_k\|}{\|r_0\|} = 10^{-8}$	$\frac{\|r_k\|}{\|r_0\|} = 10^{-12}$
	1/2	58/0.853	78/0.789	98/0.754
	1/3	164/0.945	245/0.927	365/0.927
	1/4	229/0.960	389/0.953	520/0.948
	1/6	350/0.974	669/0.973	1033/0.973
	1/8	395/0.977	805/0.977	1168/0.977
	1/12	529/0.982	1083/0.983	1682/0.983

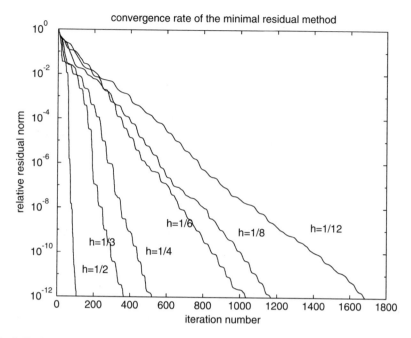

Fig. 9.10 Convergence rates of the MINRES method applied to the whole saddle-point problem

9.4   Iterative Solution of Schur Complement Systems

Because the matrix block A in (9.5) is block diagonal, the Schur complement method based on the elimination of these unknowns and subsequent iterative solution of Schur complement systems can be a very efficient technique. In particular, if we consider the matrix $\mathbb{A}$ in the form

$$\mathbb{A} = \begin{pmatrix} A & B_1 & B_2 & B_3 \\ B_1^T & 0 & 0 & 0 \\ B_2^T & 0 & 0 & 0 \\ B_3^T & 0 & 0 & 0 \end{pmatrix}, \tag{9.14}$$

then the negative of the Schur complement matrix $\mathbb{A}/A$ in the system (4.3) for the unknown $y = (y_1^T\ y_2^T\ y_3^T)^T$ is given as

$$-\mathbb{A}/A = \begin{pmatrix} B_1^T \\ B_2^T \\ B_3^T \end{pmatrix} A^{-1} \begin{pmatrix} B_1 & B_2 & B_3 \end{pmatrix} = \begin{pmatrix} A_{11} & A_{12} & A_{13} \\ A_{12}^T & A_{22} & A_{23} \\ A_{13}^T & A_{23}^T & A_{33} \end{pmatrix}, \tag{9.15}$$

where $A_{11} \in \mathcal{R}^{ne,ne}$, $A_{12} \in \mathcal{R}^{ne,nif}$, $A_{13} \in \mathcal{R}^{ne,nnc}$, $A_{22} \in \mathcal{R}^{nif,nif}$, $A_{23} \in \mathcal{R}^{nif,nnc}$, and $A_{33} \in \mathcal{R}^{nnc,nnc}$. The number of nonzero elements in the matrix $-\mathbb{A}/A$ is analyzed in Lemma 2.2 of [59]. The structure of nonzero elements of the Schur complement matrix (9.15) for the model problem with the saddle-point matrix $\mathbb{A}/A$ from Fig. 9.9 is given in Fig. 9.11. It was shown in [59] that while the Schur complement matrix (9.15) is formed by elimination of the velocity unknowns x, the second and third Schur complement system can be obtained by elimination of the pressure unknowns y_1 and of the unknowns y_3 corresponding to Lagrange multipliers. In addition, the second Schur complement matrix

$$(-\mathbb{A}/A)/A_{11} = \begin{pmatrix} A_{22} & A_{23} \\ A_{23}^T & A_{33} \end{pmatrix} - \begin{pmatrix} A_{12}^T \\ A_{13}^T \end{pmatrix} A_{11}^{-1} \begin{pmatrix} A_{12} & A_{13} \end{pmatrix} = \begin{pmatrix} B_{11} & B_{12} \\ B_{12}^T & B_{22} \end{pmatrix}, \qquad (9.16)$$

where $B_{11} \in \mathcal{R}^{nif,nif}$, $B_{12} \in \mathcal{R}^{nif,nnc}$, and $B_{22} \in \mathcal{R}^{nnc,nnc}$, is sparse, and it can be formed without any additional fill-in. For details, see the proof of Theorem 2.1

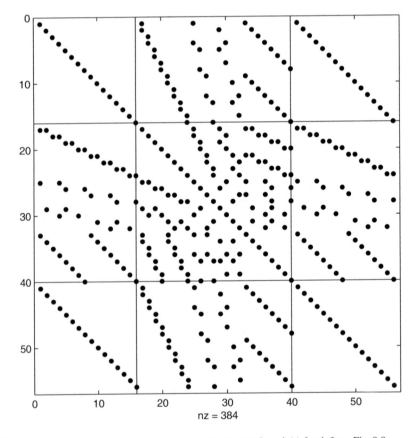

Fig. 9.11 Nonzero elements of the Schur complement matrix $-\mathbb{A}/A$ for $\mathbb{A}$ from Fig. 9.9

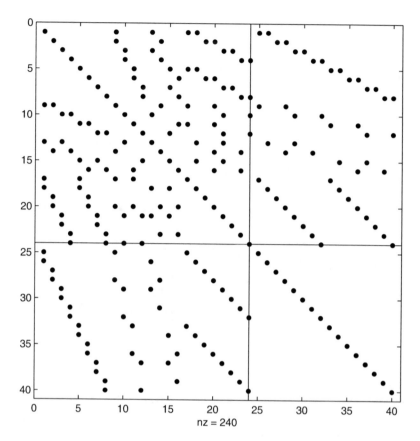

Fig. 9.12 Nonzero elements of the Schur complement matrix $(-\mathbb{A}/A)/A_{11}$ for $\mathbb{A}$ from Fig. 9.9

in [59]. The structure of nonzero elements of the Schur complement matrix (9.16) for our model problem is given in Fig. 9.12. The third Schur complement matrix given as $((-\mathbb{A}/A)/A_{11})/B_{22} = B_{11} - B_{12}B_{22}^{-1}B_{12}^T \in \mathcal{R}^{nif,nif}$ remains sparse, and it can be also formed from the second Schur complement matrix (9.16) without additional fill-in. For the proof we refer to Theorem 2.2 in [59]. In Table 9.4 we give the discretization parameters h, ne, nif, nnc, and dimension $m + n$ of the saddle-point matrix $\mathbb{A}$ for our model problem with the cubic domain. The corresponding dimensions of the Schur complement matrices $-\mathbb{A}/A$, $(-\mathbb{A}/A)/A_{11}$ and $((-\mathbb{A}/A)/A_{11})/B_{22}$ are given in Table 9.5.

The solution process can be divided into the following three main steps:

- Subsequent reduction of the saddle-point matrix $\mathbb{A}$ to the Schur complement matrices

$$\mathbb{A} \rightarrow -\mathbb{A}/A \rightarrow (-\mathbb{A}/A)/A_{11} \rightarrow ((-\mathbb{A}/A)/A_{11})/B_{22}.$$

Table 9.4 Discretization parameters h, ne, nif, nnc, and dimension $m + n$ of the saddle-point matrix $\mathbb{A}$

Mesh size h	Discretization parameters			$m + n$
	ne	nif	nnc	
1/5	250	525	199	2224
1/10	2000	4600	796	17396
1/15	6750	15,975	1791	58,266
1/20	16,000	38,400	3184	137,584
1/30	54,000	131,400	7164	462,564
1/40	128,000	313,600	12,736	1,094,336

Table 9.5 Dimension of the saddle-point matrix $\mathbb{A}$ and dimensions of the Schur complement matrices $-\mathbb{A}/A$, $(-\mathbb{A}/A)/A_{11}$ and $((-\mathbb{A}/A)/A_{11})/B_{22}$

Mesh size h	Matrix dimension			
	$\mathbb{A}$	$-\mathbb{A}/A$	$(-\mathbb{A}/A)/A_{11}$	$((-\mathbb{A}/A)/A_{11})/B_{22}$
1/5	2224	974	724	525
1/10	17,396	7396	5396	4600
1/15	58,266	24,516	17,766	15,975
1/20	137,584	57,584	41,584	38,400
1/30	462,564	192,564	138,564	131,400
1/40	1,094,336	454,336	326,336	313,600

- Iterative solution of the Schur complement system with the matrix $-((\mathbb{A}/A)/A_{11})/B_{22}$ for the unknown vector y_2.
- Block back-substitution process for computing the unknown vectors y_3, y_1, and x using the information from the Schur complement reduction.

It is clear that if A is symmetric positive definite and B has full-column rank, then all three Schur complement matrices $-\mathbb{A}/A$, $(-\mathbb{A}/A)/A_{11}$, and $((-\mathbb{A}/A)/A_{11})/B_{22}$ are symmetric positive definite. Therefore, the methods such as CG and MINRES can be applied. Since there holds the peak/plateau relation (6.33) showing essentially that there is no significant difference in the convergence of the residual norm between these two methods, we consider only the residual norm minimizing MINRES method here. Indeed, it follows from (6.34) that the convergence of the MINRES method depends on the eigenvalue distribution and the relative residual norm of MINRES can be estimated via the best minimal polynomial approximation on the spectrum of the matrices $-\mathbb{A}/A$, $(-\mathbb{A}/A)/A_{11}$, and $((-\mathbb{A}/A)/A_{11})/B_{22}$. The inclusion set for the first Schur complement matrix $-\mathbb{A}/A$ is given in (5.11). For the proof and other details, see Theorem 4.1 in [59]. Moreover, it was shown in Theorems 4.2–4.3 of [59] that the spectral properties of the second and third Schur complement matrices $(-\mathbb{A}/A)/A_{11}$ and $((-\mathbb{A}/A)/A_{11})/B_{22}$ do not deteriorate during the Schur complement reduction due to

$$\mathrm{sp}(((-\mathbb{A}/A)/A_{11})/B_{22}) \subset \mathrm{sp}((-\mathbb{A}/A)/A_{11}) \subset \mathrm{sp}(-\mathbb{A}/A). \tag{9.17}$$

Considering the inclusion set (9.6) for the spectrum of A, the inclusion set for the singular values of B given by (9.7) and the result (5.11), we get the inclusion set for the spectrum of $-\mathbb{A}/A$ in the form

$$\text{sp}(-\mathbb{A}/A) \subset [\mu_2^{-1}\mu_3^2 h^3, \mu_1^{-1}\mu_4^2 h],$$

whereas in the case of diagonal scaling (9.10), we get

$$\text{sp}(-\mathbb{A}/A) \subset [\mu_2^{-1}\mu_3^2 h^2, \mu_1^{-1}\mu_4^2]. \tag{9.18}$$

The bound for the condition number of $-\mathbb{A}/A$ is in both cases the same

$$\kappa(-\mathbb{A}/A) \leq \frac{\mu_4^2 \mu_2}{\mu_3^2 \mu_1 h^2}. \tag{9.19}$$

If we consider the bounds (6.34) and (6.29) together with (9.17), then we get the bound for the relative residual norm of the MINRES method applied either to the first, to the second, or to the third Schur complement system

$$\frac{\|r_k\|}{\|r_0\|} \leq 2 \left(\frac{1 - 1/\sqrt{\kappa(-\mathbb{A}/A)}}{1 + 1/\sqrt{\kappa(-\mathbb{A}/A)}} \right)^k. \tag{9.20}$$

Taking into account (9.19), it follows then for the relative residual norm of MINRES applied to all three Schur complement systems that

$$\frac{\|r_k\|}{\|r_0\|} \leq 2 \left(\frac{1 - \frac{\mu_3}{\mu_4}\sqrt{\frac{\mu_1}{\mu_2}}h}{1 + \frac{\mu_3}{\mu_4}\sqrt{\frac{\mu_1}{\mu_2}}h} \right)^k. \tag{9.21}$$

Based on (9.21), the asymptotic convergence factor of MINRES can be bounded as

$$\lim_{k \to +\infty} \left(\frac{\|r_k\|}{\|r_0\|} \right)^{\frac{1}{k}} \leq 1 - \xi_3 h, \tag{9.22}$$

where ξ_3 is a positive constant that depends only on the constants μ_1, μ_2, μ_3, and μ_4. Note that for the Schur complement approach, we have obtained again the bounds that depend linearly on the discretization parameter h, and they are the same for both the unscaled case with the original matrix $\mathbb{A}$ and the scaled case with (9.10).

We again illustrate our theoretical results on the model problem with a cubic domain. The computed inclusion sets for the spectrum of the Schur complement matrices $-\mathbb{A}/A$, $(-\mathbb{A}/A)/A_{11}$, and $((-\mathbb{A}/A)/A_{11})/B_{22}$ are shown in Table 9.6. It is easy to see that the inclusion set for the spectrum of $-\mathbb{A}/A$ is in a good agreement with the values predicted by (5.11). We can also see that the extremal eigenvalues of $(-\mathbb{A}/A)/A_{11}$ are bounded by the extremal eigenvalues of $-\mathbb{A}/A$. Similarly,

Table 9.6 Inclusion sets for the spectrum of the Schur complement matrices $-\mathbb{A}/A$, $(-\mathbb{A}/A)/A_{11}$ and $((-\mathbb{A}/A)/A_{11})/B_{22}$

	Inclusion sets for spectrum of matrices		
h	$-\mathbb{A}/A$	$(-\mathbb{A}/A)/A_{11}$	$((-\mathbb{A}/A)/A_{11})/B_{22}$
1/5	[0.10e0, 0.34e4]	[0.11e0, 0.12e4]	[0.2e0, 0.12e4]
1/10	[0.16e-1, 0.17e4]	[0.22e-1, 0.60e3]	[0.26e-1, 0.60e3]
1/15	[0.52e-2, 0.12e4]	[0.72e-2, 0.40e3]	[0.80e-2, 0.40e3]
1/20	[0.23e-2, 0.87e3]	[0.32e-2, 0.30e3]	[0.34e-2, 0.30e3]
1/30	[0.70e-3, 0.58e3]	[0.98e-3, 0.20e3]	[0.10e-2, 0.20e3]
1/40	[0.30e-3, 0.43e3]	[0.42e-3, 0.15e3]	[0.43e-3, 0.15e3]

Table 9.7 Number of iterations steps of MINRES applied to the Schur complement systems with $-\mathbb{A}/A$, $(-\mathbb{A}/A)/A_{11}$ and $((-\mathbb{A}/A)/A_{11})/B_{22}$

	Number of iteration steps		
h	$-\mathbb{A}/A$	$(-\mathbb{A}/A)/A_{11}$	$((-\mathbb{A}/A)/A_{11})/B_{22}$
1/5	82	47	43
1/10	154	87	80
1/15	223	126	118
1/20	288	164	155
1/25	353	199	192
1/30	418	234	228
1/35	482	269	263
1/40	546	303	298

the extremal eigenvalues of $((-\mathbb{A}/A)/A_{11})/B_{22}$ are bounded by the extremal eigenvalues of $(-\mathbb{A}/A)/A_{11}$. In Table 9.7 we give the number of iteration steps of the MINRES method applied to the resulting three Schur complement systems with the initial guess set to zero vector and the relative residual norm $\frac{\|r_k\|}{\|r_0\|} = 10^{-8}$ used as a stopping criterion. The dependence of the number of iteration steps in Table 9.7 corresponds well to the asymptotic behavior with the linear dependence on h. The same conclusion can be clearly drawn also from Fig. 9.13, where we show the convergence rate of MINRES applied to the third Schur complement system for several values of the discretization parameter h.

9.5 Iterative Solution of Systems Projected onto Null-Spaces

Since the off-diagonal matrix block $B = (B_1\ B_2\ B_3)$ in our application has a simple structure, another reasonable approach for solving the saddle-point problem (2.20) is to construct a basis of $\mathcal{N}(B^T)$ or a basis of the null-space of a certain sub-block of B^T, and then to solve a projected system using some iterative method.

First, we will discuss the standard null-space method based on a basis of the whole off-diagonal block B^T. A model example of the off-diagonal block B^T is

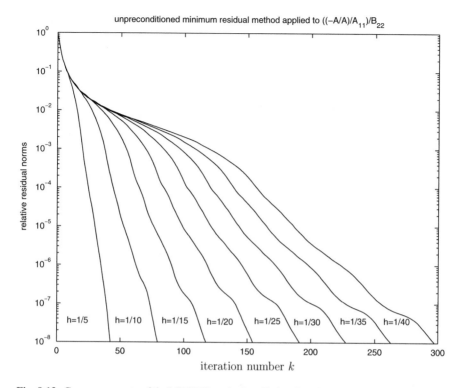

Fig. 9.13 Convergence rate of the MINRES method applied to the third Schur complement system

shown in Fig. 9.14. It was shown in [4] that B^T represents an incomplete incidence matrix of a certain graph, and one way how to compute the null-space basis Z satisfying $B^T Z = 0$ is to take the incidence vectors of some of its cycles. The construction of the so-called fundamental cycle null-space basis is based on the algorithm that finds the shortest path spanning tree, and forms cycles using different non-tree edges in the graph to ensure the linear independence of the resulting basis. For details we refer to Subsection 2.1. in [4]. Note that the vectors in the fundamental cycle null-space basis are linearly independent but they are not orthogonal. It was also shown in Theorems 2.2–2.3 of [4] that the singular values of the matrix Z with the fundamental null-space basis as its column vectors are included in the interval

$$\mathrm{sv}(Z) \subset \left[1, \frac{\mu_5}{h^2}\right], \tag{9.23}$$

where μ_5 is a positive constant such that the term $\mu_5 h^{-2}$ in (9.23) gives in fact a measure for the longest cycle in the associated graph [4]. Using the inclusion set (9.6) for the spectrum of A, the inclusion set (9.23) for the singular values of Z,

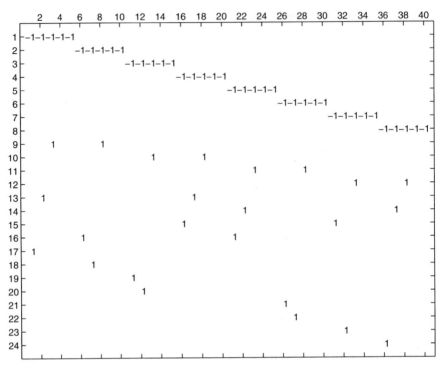

Fig. 9.14 Example of the off-diagonal block B^T in a simple model problem

and the result (5.22), we get the inclusion set for the matrix $Z^T A Z$ in the form

$$\text{sp}(Z^T A Z) \subset \left[\frac{\mu_1}{h}, \frac{\mu_2 \mu_5^2}{h^5} \right]. \tag{9.24}$$

If we consider the diagonal scaling (9.10), then we get the inclusion set

$$\text{sp}(Z^T (hA)Z) \subset \left[\mu_1, \frac{\mu_2 \mu_5^2}{h^4} \right]. \tag{9.25}$$

However, the bound for the condition number is in both cases

$$\kappa(Z^T A Z) \leq \frac{\mu_2 \mu_5^2}{\mu_1 h^4}. \tag{9.26}$$

If we consider again the bounds (6.34) and (6.29), then we get the bound for the relative residual norm of the MINRES method applied to the projected system (4.14) in the form

$$\frac{\|r_k\|}{\|r_0\|} \leq 2 \left(\frac{1 - 1/\sqrt{\kappa(Z^T A Z)}}{1 + 1/\sqrt{\kappa(Z^T A Z)}} \right)^k . \tag{9.27}$$

From (9.26) and (9.27) for the relative residual norm of MINRES applied to the projected system, it follows that

$$\frac{\|r_k\|}{\|r_0\|} \leq 2 \left(\frac{1 - \frac{1}{\mu_5} \sqrt{\frac{\mu_1}{\mu_2}} h^2}{1 + \frac{1}{\mu_5} \sqrt{\frac{\mu_1}{\mu_2}} h^2} \right)^k . \tag{9.28}$$

This leads to the bound for the asymptotic convergence factor of MINRES

$$\lim_{k \to +\infty} \left(\frac{\|r_k\|}{\|r_0\|} \right)^{\frac{1}{k}} \leq 1 - \xi_4 h^2, \tag{9.29}$$

where the positive constant ξ_4 depends only on the constants μ_1, μ_2, and μ_5. Note that due to the potential ill-conditioning of the fundamental cycle null-space basis Z, the bound (9.29) depends quadratically on the discretization parameter h, but as it is indicated later in experiments, it seems to be quite pessimistic.

The solution process can be divided into the following four steps:

- Construction of an easy-to-compute but not necessarily orthonormal basis Z of $\mathcal{N}(B^T)$.
- Construction of a particular solution $\hat{x}$ to underdetermined system $B^T \hat{x} = d$.
- Iterative solution of the projected system $Z^T A Z \tilde{x} = Z^T (c - A\hat{x})$.
- Back-substitution to compute the unknown vector y so that $By = c - Ax$, where $x = \hat{x} + Z\tilde{x}$.

Although the off-diagonal matrix B can be ill-conditioned, its submatrix $(B_2 \ B_3)$ has orthogonal columns. See, e.g., the model example in Fig. 9.14. Therefore, it is much easier to construct a null-space basis only for $(B_2 \ B_3)^T$ than for the whole matrix B^T. In contrast to the previous approach, this basis can be explicitly constructed orthogonal, and thus its condition number does not depend on the discretization parameter h. The null-space matrix Z for our model example is given in Fig. 9.15. Note that, although we are splitting the potentially ill-conditioned matrix B into two submatrices B_1 and $(B_2 \ B_3)$ with orthogonal columns, the spectrum of the saddle-point matrix $\mathbb{A}$ projected on the null-space of $(B_2 \ B_3)^T$ is dependent on the discretization, as it will be visible in (9.32).

Fig. 9.15 Null-space basis of
the off-diagonal blocks
$(B_2\ B_3)^T$ in the model
example from Fig. 9.14

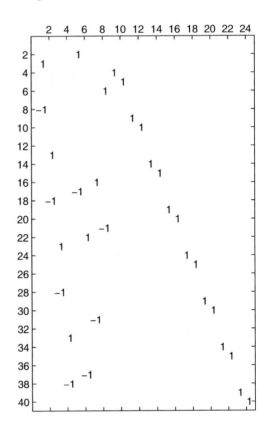

We consider the solution approach divided into the following main steps:

- Explicit construction of the orthogonal basis Z of $\mathcal{N}((B_2\ B_3)^T)$.
- Construction of some particular solution to the underdetermined system
$\begin{pmatrix} B_2^T \\ B_3^T \end{pmatrix} \hat{x} = \begin{pmatrix} d_2 \\ d_3 \end{pmatrix}$.
- Iterative solution of the projected system

$$\begin{pmatrix} Z^T A Z & Z^T B_1 \\ B_1^T Z & 0 \end{pmatrix} \begin{pmatrix} \tilde{x} \\ y_1 \end{pmatrix} = \begin{pmatrix} Z^T (c - A\hat{x}) \\ d_1 - B_1^T \hat{x} \end{pmatrix}. \qquad (9.30)$$

- Back-substitution to compute the unknown vector $\begin{pmatrix} y_2 \\ y_3 \end{pmatrix}$ so that

$$(B_2\ B_3) \begin{pmatrix} y_2 \\ y_3 \end{pmatrix} = c - Ax - B_1 y_1,$$

where $x = \hat{x} + Z\tilde{x}$.

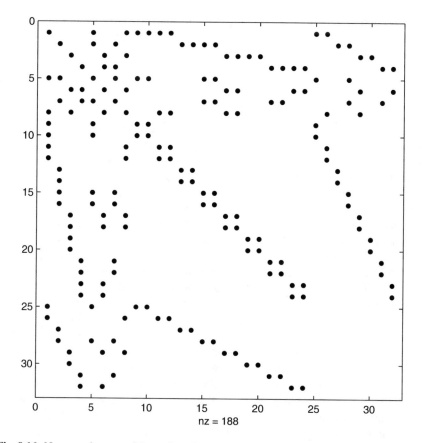

Fig. 9.16 Nonzero elements of the projected matrix (9.30) in our simple model example

The projected system (9.30) is again a saddle-point problem. The structure of its nonzero elements is shown in Fig. 9.16. It was shown in Lemma 3.2 of [4] that if we consider the diagonal scaling (9.10), then there exist positive constants μ_6 and μ_7 such that singular values of the matrix block $Z^T B_1$ belong to the interval

$$\text{sv}(Z^T B_1) \subset [\mu_6 h, \mu_7]. \tag{9.31}$$

Using $\text{sp}(hA) \subset [\mu_1, \mu_2]$ and the result (3.16), we get the inclusion set for the spectrum of the projected matrix in (9.30) in the form

$$\left[\frac{1}{2}(\mu_1 - \sqrt{\mu_1^2 + 4\mu_7^2}), -\frac{\mu_6^2}{\mu_2} h^2\right] \cup \left[\mu_1, \frac{1}{2}(\mu_2 + \sqrt{\mu_2^2 + 4\mu_7^2})\right]. \tag{9.32}$$

Thus we obtain the result that is completely analogous to the case of the inclusion set (9.11). So, if we apply the MINRES method to the projected symmetric indefinite system (9.30), then the bound for its asymptotic convergence factor is given as

$$\lim_{k \to \infty} \left(\frac{\|r_k\|}{\|r_0\|} \right)^{\frac{1}{k}} \leq 1 - \xi_4 h, \tag{9.33}$$

where the positive constant ξ_4 depends only on the constants μ_1, μ_2, μ_6, and μ_7. Indeed, if we use the diagonal scaling (9.10) together with this variant of the null-space method, then the asymptotic rate of convergence of the MINRES method depends at most linearly on the parameter h. For details we refer to [4].

In the remaining text, we illustrate our theoretical results on the model problem with a cubic domain. In Table 9.8 we give discretization parameters h, ne, nif, and nnc and dimensions of null-spaces $nz1$ and $nz2$ for several values of discretization parameter h. Note that the dimension of the null-space of B^T is equal to $nz1 = 4 \cdot ne - nif - nnc$, and the dimension of $(B_2 \ B_3)^T$ is equal to $nz2 = 4 \cdot ne - nif - nnc$. In Table 9.9 we give the number of nonzero entries in the fundamental cycle null-space basis of $\mathcal{N}(B^T)$ (denoted as $nnz1$) and in the orthogonal basis of $\mathcal{N}((B_2 \ B_3)^T)$ (denoted as $nnz2$). In Table 9.9 we also consider the number of iteration steps of the MINRES method applied to the resulting projected system (4.14) with the initial guess to the unknown x set to zero and with the relative residual norm $\frac{\|r_k\|}{\|r_0\|} = 10^{-8}$ used as a stopping criterion (denoted as $nit1$). We also show the number of iteration steps of MINRES applied to the projected system (9.30) and preconditioned with the constraint preconditioner (7.31) with the zero initial guess and with the same stopping criterion (denoted as $nit2$). This approach essentially corresponds to the unpreconditioned null-space method applied to the system (9.30). The dependence of the number of iteration steps in Table 9.9 in both cases corresponds well to the bounds with the linear dependence on h, and it seems that the bound for the fundamental cycle basis is rather pessimistic. In Figs. 9.17 and 9.18, we show the corresponding convergence rates of MINRES for several values of the discretization parameter h.

Table 9.8 Discretization parameters h, ne, nif, and nnc and dimensions of null-spaces $nz1$ and $nz2$

Discretization parameters				Dimension of null-spaces	
h	ne	nif	nnc	$nz1$	$nz2$
1/5	250	525	100	375	625
1/10	2000	4600	400	3000	5000
1/15	6750	15,975	900	10,125	16,875
1/20	16,000	38,400	1600	24,000	40,000
1/25	31,250	75,625	2500	46,875	78,125
1/30	54,000	131,400	3600	81,000	135,000
1/35	87,750	209,475	4900	138,625	226,375
1/40	128,000	313,600	6400	192,000	320,000

Table 9.9 Number of nonzero entries in the fundamental cycle null-space basis of $\mathcal{N}(B^T)$ and in the orthogonal basis of $\mathcal{N}((B_2 \, B_3)^T)$ Number of iterations steps of unpreconditioned and preconditioned MINRES applied to the projected system (4.14) and (9.30), respectively

	Memory requirements		Iteration count	
h	$nnz1$	$nnz2$	$nit1$	$nit2$
1/5	3360	14,375	71	35
1/10	47,120	123,000	163	64
1/15	226,780	424,125	252	93
1/20	697,840	1,016,000	346	118
1/25	1,675,800	1,996,875	438	145
1/30	3,436,160	3,465,000	523	174
1/35	6,314,420	5,518,625	596	204
1/40	10,706,080	8,256,000	670	230

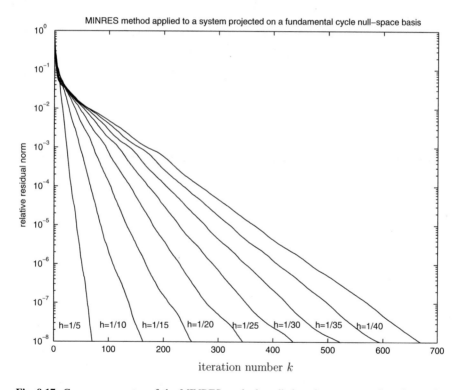

Fig. 9.17 Convergence rates of the MINRES method applied to the system projected onto the null-space of B^T

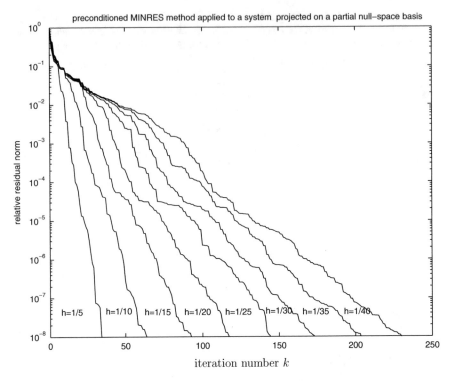

Fig. 9.18 Convergence rates of the preconditioned MINRES method applied to the system projected onto the null-space of $(B_2 \; B_3)^T$

Bibliography

1. M. Arioli, The use of QR factorization in sparse quadratic programming and backward error issues. SIAM J. Matrix Anal. Appl. **21**, 825–839 (2000)
2. M. Arioli, L. Baldini, A backward error analysis of a null space algorithm in sparse quadratic programming. SIAM J. Matrix Anal. Appl. **23**, 425–442 (2001)
3. M. Arioli, G. Manzini, A network programming approach in solving Darcy's equations by mixed finite-element methods. ETNA **22**, 41–70 (2006)
4. M. Arioli, J. Maryška, M. Rozložník, M. Tůma, Dual variable methods for mixed-hybrid finite element approximation of the potential fluid flow problem in porous media. ETNA **22**, 17–40 (2006)
5. D.N. Arnold, R.S. Falk, R. Winther, Preconditioning in $H(\text{div})$ and applications. Math. Comput. **66**(219), 957–984 (1997)
6. O. Axelsson, J. Karátson, Equivalent operator preconditioning for elliptic problems. Numer. Algorithms **50**(3), 297–380 (2009)
7. Z.Z. Bai, Solutions of Linear Systems of Block Two-by-Two Structures. http://lsec.cc.ac.cn/~bzz/Public/psfiles/slice22a.pdf
8. Z.Z. Bai, G.H. Golub, M.K. Ng, Hermitian and skew-Hermitian splitting methods for non-Hermitian positive definite linear systems. SIAM J. Matrix Anal. Appl. **24**, 603–626 (2003)
9. P. Bastian, M. Blatt, R. Scheichl, Algebraic multigrid for discontinuous Galerkin discretizations of heterogeneous elliptic problems. Numer. Linear Algebra Appl. **19**(2), 367–388 (2012)
10. M. Benzi, Preconditioning techniques for large linear systems: a survey. J. Comput. Phys. **182**, 418–477 (2002)
11. M. Benzi, V. Simoncini, On the eigenvalues of a class of saddle point matrices. Numer. Math. **103**, 173–196 (2006)
12. M. Benzi, G.H. Golub, J. Liesen, Numerical solution of saddle point problems. Acta Numerica **14**, 1–137 (2005)
13. L. Bergamaschi, J. Gondzio, G. Zilli, Preconditioning indefinite systems in interior point methods for optimization. Comput. Optim. Appl. **28**(2), 149–171 (2004)
14. A. Björck. *Numerical Methods for Least Squares Problems* (SIAM, Philadelphia, 1996)
15. D. Braess, R. Sarazin, An efficient smoother for the Stokes problem. Appl. Numer. Math. **23**, 3–20 (1997)
16. J.H. Bramble, J.E. Pasciak, A preconditioning technique for indefinite systems resulting from mixed approximations of elliptic problems. Math. Comput. **50**, 1–17 (1988)
17. F. Brezzi, M. Fortin, *Mixed and Hybrid Finite Element Methods* (Springer, New York, 1991)
18. Y. Chabrillac, J.P. Crouzeix, Definiteness and semidefiniteness of quadratic forms revisited. Linear Algebra Appl. **63**(1), 283–292 (1984)

© Springer Nature Switzerland AG 2018

M. Rozložník, *Saddle-Point Problems and Their Iterative Solution*, Nečas Center Series, https://doi.org/10.1007/978-3-030-01431-5

19. D. Drzisga, L. John, U. Rüde, B. Wohlmuth, W. Zulehner, On the analysis of block smoothers for saddle point problems. SIAM J. Matrix Anal. Appl. **39**(2), 932–960 (2018)

20. C. Durazzi, V. Ruggiero, Indefinitely preconditioned conjugate gradient method for large sparse equality and inequality constrained quadratic problems. Numer. Linear Algebra Appl. **10**(8), 673–688 (2002)

21. H.C. Elman, Multigrid and Krylov subspace methods for the discrete Stokes equations. Int. J. Numer. Meth. Fluids **22**, 755–770 (1996)

22. H.C. Elman, G.H. Golub, Inexact and preconditioned Uzawa algorithms for saddle point problems. SIAM J. Numer. Anal. **31**(6), 1645–1661 (1994)

23. H.C. Elman, D.J. Silvester, A.J. Wathen, Iterative methods for problems in computational fluid dynamics, in *Iterative Methods in Scientific Computing*, ed. by R.H. Chan, C.T. Chan, G.H. Golub (Springer, Singapore, 1997), pp. 271–327

24. H. Elman, D.J. Silvester, A.J. Wathen, Block preconditioners for the discrete incompressible Navier-Stokes equations. Int. J. Numer. Meth. Fluids **40**, 333–344 (2002)

25. H. Elman, D.J. Silvester, A.J. Wathen, *Finite Elements and Fast Iterative Solvers with Applications in Incompressible Fluid Dynamics* (Oxford University Press, Oxford, 2005)

26. R. Estrin, C. Greif, On nonsingular saddle-point systems with a maximally rank deficient leading block. SIAM J. Matrix Anal. Appl. **36**(2), 367–384 (2015)

27. R. Estrin, C. Greif, Towards an optimal condition number of certain augmented Lagrangian-type saddle-point matrices. Numer. Linear Algebra Appl. **23**(4), 693–705 (2016)

28. V. Faber, T.A. Manteuffel, S.V. Parter, On the theory of equivalent operators and application to the numerical solution of uniformly elliptic partial differential equations. Adv. Appl. Math. **11**(2), 109–163 (1990)

29. B. Fischer, A. Ramage, D.J. Silvester, A.J. Wathen, Minimum residual methods for augmented systems. BIT **38**, 527–543 (1998)

30. R.W. Freund, N.M. Nachtigal, QMR: a quasi-minimal residual method for non-Hermitian linear systems. Numer. Math. **60**, 315–339 (1991)

31. R.W. Freund, N.M. Nachtigal, Software for simplified Lanczos and QMR algorithms. Appl. Numer. Math. **19**, 319–341 (1995)

32. R. Freund, G.H. Golub, N.M. Nachtigal, Iterative solution of linear systems. Acta Numerica **1**, 1–44 (1992)

33. J.F. Gerbeau, C. Farhat, CME358: The Finite Element Method for Fluid Mechanics. Stanford University, Spring 2009. http://www.stanford.edu/class/cme358/

34. T. Gergelits, Z. Strakoš, Composite convergence bounds based on Chebyshev polynomials and finite precision conjugate gradient computations. Numer. Algorithms **65**(4), 759–782 (2014)

35. G. Golub, C. Greif, On solving block-structured indefinite linear systems. SIAM J. Sci. Comput. **24**(6), 2076–2092 (2003)

36. N.I.M. Gould, V. Simoncini, Spectral analysis of saddle point matrices with indefinite leading blocks. SIAM J. Matrix Anal. Appl. **31**(3), 1152–1171 (2009)

37. A. Greenbaum, *Iterative Methods for Solving Linear Systems* (SIAM, Philadelphia, 1997)

38. A. Greenbaum, V. Pták, Z. Strakoš, Any convergence curve is possible for GMRES. SIAM J. Matrix Anal. Appl. **17**, 465–470 (1996)

39. M.R. Hestenes, E. Stiefel, Methods of conjugate gradients for solving linear systems. J. Res. Nat. Bur. Stand. **49**, 409–436 (1952)

40. N.J. Higham, *Accuracy and Stability of Numerical Algorithms*, 2nd edn. (SIAM, Philadelphia, 2002)

41. R. Hiptmair, Operator preconditioning. Comput. Math. Appl. **52**(5), 699–706 (2006)

42. R.A. Horn, C.R. Johnson, *Matrix Analysis*, 2nd edn. (Cambridge University Press, New York, 2012)

43. J. Hrnčíř, I. Pultarová, Z. Strakoš, Decomposition into subspaces and operator preconditioning I: abstract framework. Preprint CNMM/2017/06, Prague (2017)

44. I.C.F. Ipsen, A note on preconditioning non-symmetric matrices. SIAM J. Sci. Comput. **23**(3), 1050–1051 (2001)

45. P. Jiránek, M. Rozložník, Maximum attainable accuracy of inexact saddle point solvers. SIAM J. Matrix Anal. Appl. **29**(4), 1297–1321 (2008)
46. P. Jiránek, M. Rozložník, Limiting accuracy of segregated solution methods for nonsymmetric saddle point problems. J. Comput. Appl. Math. **215**, 28–37 (2008)
47. E.F. Kaasschieter, A.J.M. Huijben, Mixed hybrid finite elements and streamline computation for the potential flow problem. Numer. Method Partial Differ. Equ. **8**(3), 221–266 (1992)
48. A. Klawonn, Block-triangular preconditioners for saddle point problems with a penalty term. SIAM J. Sci. Comput. **19**(1), 172–184 (1998)
49. A. Klawonn, An optimal preconditioner for a class of saddle point problems with a penalty term. SIAM J. Sci. Comput. **19**(2), 540–552 (1998)
50. M. Kroupa, J. Mužák, J. Trojáček, Remediation of former uranium in-situ leaching area at Stráž pod Ralskem – Hamr na Jezeře, Czech Republic. The Uranium Mining and Remediation Exchange Group (UMREG) and other uranium production cycle technical meetings – Selected papers 2012–2015. IAEA-TECDOC, Vienna (in preparation)
51. J. Liesen, Z. Strakoš, Convergence of GMRES for tridiagonal toeplitz matrices. SIAM J. Matrix Anal. Appl. **26**, 233–251 (2004)
52. J. Liesen, Z. Strakoš, GMRES convergence analysis for a convection-diffusion model problem. SIAM J. Sci. Comput. **26**, 1989–2009 (2005)
53. J. Liesen, B. Parlett, On nonsymmetric saddle point matrices that allow conjugate gradient iterations. Numer. Math. **108**, 605–624 (2008)
54. J. Liesen, Z. Strakoš, *Krylov Subspace Methods, Principles and Analysis* (Oxford University Press, Oxford, 2013)
55. L. Lukšan, J. Vlček, Indefinitely preconditioned inexact Newton method for large sparse equality constrained non-linear programming problems. Numer. Linear Algebra Appl. **5**(3), 1099–1506 (1999)
56. J. Málek, Z. Strakoš, *Preconditioning and the Conjugate Gradient Method in the Context of Solving PDEs*. SIAM Spotlight Series (SIAM, Philadelphia, 2015)
57. J. Maryška, M. Rozložník, M. Tůma, Mixed hybrid finite element approximation of the potential fluid flow problem. J. Comput. Appl. Math. **63**, 383–392 (1995)
58. J. Maryška, M. Rozložník, M. Tůma, The potential fluid flow problem and the convergence rate of the minimal residual method. Numer. Linear Algebra Appl. **3**, 525–542 (1996)
59. J. Maryška, M. Rozložník, M. Tůma, Schur complement systems in the mixed-hybrid finite element approximation of the potential fluid flow problem. SIAM J. Sci. Comput. **22**, 704–723 (2000)
60. J. Maryška, M. Rozložník, M. Tůma, Schur complement reduction in the mixed-hybrid approximation of Darcy's law: rounding error analysis. J. Comput. Appl. Math. **117**, 159–173 (2000)
61. M.F. Murphy, G.H. Golub, A.J. Wathen, A note on preconditioning for indefinite linear systems. SIAM J. Sci. Comput. **21**, 1969–1972 (2000)
62. J. Nocedal, S.J. Wright, *Numerical Optimization*, 2nd edn. (Springer, New York, 2006)
63. D. Orban, M. Arioli, *Iterative Solution of Symmetric Quasi-Definite Linear Systems*. SIAM Spotlight Series (SIAM, Philadelphia, 2017)
64. C.C. Paige, M.A. Saunders, Solution of sparse indefinite systems of linear equations. SIAM J. Numer. Anal. **12**, 617–629 (1975)
65. I. Perugia, V. Simoncini, Block-diagonal and indefinite symmetric preconditioners for mixed finite element formulations. Numer. Linear Algebra Appl. **7**(7–8), 585–616 (2000)
66. J. Pestana, A.J. Wathen, The antitriangular factorization of saddle point matrices. SIAM J. Matrix Anal. Appl. **35**(2), 339–353 (2014)
67. J. Pestana, A.J. Wathen, Natural preconditioning and iterative methods for saddle point systems. SIAM Rev. **57**, 71–91 (2015)
68. C. Powell, D.J. Silvester, Optimal preconditioning for Raviart–Thomas mixed formulation of second-order elliptic problems. SIAM J. Matrix Anal. Appl. **25**(3), 718–738 (2003)
69. A. Quarteroni, A. Vialli, *Numerical Approximation of Partial Differential Equations* (Springer, Berlin/Heidelberg, 1994)

70. T. Rees, J. Scott, The null-space method and its relationship with matrix factorizations for sparse saddle point systems. Numer. Linear Algebra Appl. **25**(1), 1–17 (2018)
71. M. Rozložník, V. Simoncini, Krylov subspace methods for saddle point problems with indefinite preconditioning. SIAM J. Matrix Anal. Appl. **24**(2), 368–391 (2002)
72. M. Rozložník, F. Okulicka-Dłużewska, A. Smoktunowicz, Cholesky-like factorization of symmetric indefinite matrices and orthogonalization with respect to bilinear forms. SIAM J. Matrix Anal. Appl. **36**(2), 727–751 (2015)
73. T. Rusten, R. Winther, A preconditioned iterative method for saddlepoint problems. SIAM J. Matrix Anal. Appl. **13**(3), 887–904 (1992)
74. Y. Saad, *Iterative Methods for Sparse Linear Systems* (PWS Pub. Co., Boston, 1996) (2nd edition: SIAM, Philadelphia, 2003)
75. Y. Saad, M.H. Schultz, GMRES: a generalized minimal residual algorithm for solving nonsymmetric linear systems. SIAM J. Sci. Stat. Comput. **7**, 856–869 (1986)
76. D.J. Silvester, A.J. Wathen, Fast iterative solution of stabilized Stokes systems, part II: using block preconditioners. SIAM J. Numer. Anal. **31**, 1352–1367 (1994)
77. J. Stoer, Solution of large linear systems of conjugate gradient type methods, in *Mathematical Programming* (Springer, Berlin, 1983), pp. 540–565
78. J. Stoer, R. Freund, On the solution of large indefinite systems of linear equations by conjugate gradients algorithm, in *Computing Methods in Applied Sciences and Engineering V*, ed. by R. Glowinski, J.L. Lions (INRIA, North Holland, 1982), pp. 35–53
79. M. Stoll, A. Wathen, Combination preconditioning and the Bramble-Pasciak+ preconditioner. SIAM J. Matrix Anal. Appl. **30**(2), 582–608 (2008)
80. R.J. Vanderbei, Symmetric quasi-definite matrices. SIAM J. Optim. **5**(1), 100–113 (1995)
81. A.J. Wathen, Preconditioning. Acta Numerica **24**, 329–376 (2015)
82. A.J. Wathen, B. Fischer, D.J. Silvester, The convergence rate of the minimal residual method for the Stokes problem. Numer. Math. **71**, 121–134 (1995)
83. A.J. Wathen, B. Fischer, D.J. Silvester, The convergence of iterative solution methods for symmetric and indefinite linear systems, in *Numerical Analysis*, ed. by D. Griffiths, D. Higham, A.G. Watson (Longman Scientific, Harlow, 1998), pp. 230–240
84. W. Zulehner, A class of smoothers for saddle point problems. Computing **65**, 227–246 (2000)
85. W. Zulehner, Analysis of iterative methods for saddle point problems: a unified approach. Math. Comput. **71**, 479–505 (2002)
86. W. Zulehner, Nonstandard norms and robust estimates for saddle point problems. SIAM J. Matrix Anal. Appl. **32**(2), 536–560 (2011)

Index

© Springer Nature Switzerland AG 2018
M. Rozložník, *Saddle-Point Problems and Their Iterative Solution*,
Nečas Center Series, https://doi.org/10.1007/978-3-030-01431-5

Printed in the United States
By Bookmasters